Fisheries Conflicts in the North Atlantic: Problems of Management and Jurisdiction

Fisheries Conflicts in the North Atlantic: Problems of Management and Jurisdiction

Edited by Giulio Pontecorvo
with the editorial assistance of Norma Hench Hagist

Law of the Sea Institute Workshop
Hamilton, Bermuda
January 1974

Ballinger Publishing Company   •   Cambridge, Mass.
A Subsidiary of J. B. Lippincott Company

Fisheries Conflicts in the North Atlantic: Problems of Management and Jurisdiction, a workshop held in Hamilton, Bermuda, January 14-17, 1974, was fourth in a series of international workshops conducted by the Law of the Sea Institute, University of Rhode Island. Principal support for all workshops came from a special grant to the Law of the Sea Institute from the Ford Foundation. Additional support for this workshop was provided by the National Marine Fisheries Service of the National Oceanic and Atmospheric Administration, United States Department of Commerce.

The Law of the Sea Institute gratefully acknowledges support of its 1973-1974 activities by the Ford Foundation, the National Sea Grant Program and National Marine Fisheries Service of the National Oceanic and Atmospheric Administration, the United States Coast Guard, and the University of Rhode Island.

International Standard Book Number: 0-88410-020-0

Library of Congress Catalog Card Number: 74-9665

Printed in the United States of America

Library of Congress Cataloging in Publication Data
Main entry under title:

Fisheries Conflicts in the North Atlantic.

A Law of the Sea Institute workshop, held in Hamilton, Bermuda, January 14-17, 1974.
1. Fishery management—North Atlantic Ocean.
2. Fishery law and legislation—North Atlantic Ocean.
I. Pontecorvo, Giulio, 1923-   ed. II. Law of the Sea Institute.
SH328.F55   338.3'72'709214   74-9665
ISBN 0-88410-020-0

Law of the Sea Institute
Executive Board

Law of the Sea Institute

Contents

List of Figures

List of Tables

List of Abbreviations

CLS	Catch limitation scheme
EEC	European Economic Community
EFZ	Exclusive fishery zone
EZ	Economic zone
FAO	Food and Agriculture Organization [of the United Nations]
GRT	Gross Registered Tonnage
IATTC	Inter-American Tropical Tuna Commission
ICCAT	International Convention for the Conservation of Atlantic Tunas
ICES	International Council for the Exploration of the Sea
ICJ	International Court of Justice
ICNAF	International Commission for the Northwest Atlantic Fisheries
IMCO	Intergovernmental Maritime Consultative Organization
INPFC	International North Pacific Fisheries Commission
IWC	International Whaling Commission
LNG	Liquid natural gas
LOS III	Third Law of the Sea Conference [of the United Nations]
MSY	Maximum Sustainable Yield
NEAFC	North-East Atlantic Fisheries Commission
OECD	Organisation for Economic Cooperation and Development

"Fundamentally, therefore, the practical conflict has been a question of pace: preference moves more slowly than priority but, given continued coastal state growth, both eventually mean exclusive exploitation by the coastal state. Again in practice, if not in theory also, exclusive exploitation must give rise to exclusive jurisdiction. Hence, preference, priority and exclusive jurisdiction belong to a single continuum or almost an inexorable line of development.

....The hope must be, however, that exclusive jurisdiction will not be confused with exclusive use by the coastal state and that fish which could be economically harvested by others will not be allowed to die of old age simply because their capture is beyond the capacity of the coastal state."

A. Laing

Hamilton, Bermuda
January 1974

Preface

One of the intellectual glories of the scientific revolution of the seventeenth century was the integration of the basic theorems of mechanics into a system. By the latter part of the nineteenth century neoclassical economic theorists had begun to utilize the calculus and the methodological framework of mechanics in their analysis of economic problems. While this development helped unravel the classical economic mystery of price and value, the course of events in the twentieth century gradually revealed the Sisyphean character of this advance. Statics is both a useful methodology and a good starting point in the analysis of complex social problems. If we wish, however, to extend our knowledge to the point where we may effectively manage our economic and social affairs, static systems do little more than suggest the need of looking further into dynamic or disequilibrium systems.

Nowhere is this more apparent than in today's attempts to deal with the problems of international fisheries. Heretofore, the basic tool of international fisheries management has been the institution of the fisheries commission. These commissions, the product of bilateral or multilateral international agreements, have tended to base their management approach on the biological concept of the maximum sustainable yield (MSY). In turn, the MSY is just one point on a yield function that is a mathematical representation of a particular fish population under long-run equilibrium conditions.

A simple listing of the forces currently influencing events in international fisheries is sufficient to suggest both the limitations of the institution and the theoretical framework it utilizes to meet today's fisheries management problems. Examination of international fisheries indicates an increasing demand for fish and fish products. This increase in demand flows from increases in

income throughout the world as well as a realization of the need for protein
from the sea for nutritional purposes. On the supply side, we find that the
once-assumed unlimited harvest from the oceans is in reality becoming increasingly scarce. Furthermore, it appears that there is very little possibility for
significant increases in the catch of fish in the near future.

This supply-demand imbalance is compounded by rapid improve-
ments in the technology utilized in catching fish. The international fisheries are
becoming (in the face of the limited supply) more and more capital intensive.
Under these pressures the international fisheries move steadily toward an organizational structure that yields rising prices and has more and more labor and
capital pursuing the same number or fewer fish. At the very least, this is a
formula for economic failure.

The legal foundations of international fisheries are in equal disarray.
The common property status of the fish stocks has presented a fisheries management problem that is, as a generation of economists have pointed out, solvable.
Yet, to date, progress toward the necessary solution of entry limitations may be
generously described as proceeding at a snail's pace. While certain recent developments may suggest that if we are optimistic there may be some light at the
end of that legal tunnel, another legal issue—the extent of national jurisdiction—
has arisen to introduce a new set of problems and significantly raises the already
high level of uncertainty in the fisheries management problem.

Consider the following scenario as an example of what types of
problems may arise. Assume a general extension of national jurisdiction to 200
miles. This extension will effectively close the northeastern Atlantic on roughly
a line from the French-Spanish border to southern Greenland to all but littoral
states. This shutting off of the Northeast Atlantic may force a readjustment of
the current Russian and Eastern European effort in the enclosed area to the
Pacific. This, in turn, will cause a major readjustment in the western Pacific.

From an analytical point of view, undue fascination with alternative
patterns of cascading dominoes is perhaps the first stage of madness. But what
is real in this scenario is the increase in uncertainty associated with the rapidly
approaching changes in jurisdiction proposed in anticipation of the Law of the
Sea Conference.

The principal proximate cost associated with increased uncertainty
about the emerging regime of the oceans and the implications of that regime for
the distribution of wealth and income is its impact on the Law of the Sea
Conference itself. In negotiations already made unduly complex by the number
of participants, the added uncertainty makes the identification of each nation's
interest in alternative solutions that much harder. Uncertainty surrounding the
identification of national interest in the negotiations has also contributed to the
formalizing and at least, in the penultimate pre-conference stage, the rigidifying
of government negotiating positions.

It was the combination of academic concern with the complexity of the fisheries management problem, the current impasse in thinking about alternatives, and the implications of proposed jurisdictional changes that led the Law of the Sea Institute to hold this workshop on the fisheries problems of the North Atlantic. It is our hope that these papers will serve to emphasize the range and complexity of the issues involved in international fisheries management and also to act as a catalyst for further research and public discussion. This is particularly needed at a time when both these commodities are subordinated to the need for more simplistic positions that are politically negotiable in the uncertain environment.

The papers may be divided into three groups. The papers in the first group—by Alexander, Andersen, Johnston, and Koers—deal in their several ways with the North Atlantic as a region; its geographic definition, the cost of change to the indigenous cultures and peoples of the area, and the implications of change for both the legal structure and the fisheries interests represented by today's structure.

The papers in the second group—by McHugh, Christy, Laing, Schram, and Adam—develop aspects of the biological, economic, and political forces currently operating in the region, and they also explore the implications for the future of alternative legal regimes.

The final papers by Bilder, Clingan, and Seaton investigate the necessary conditions of a new legal framework if we wish the framework that will be needed to minimize conflicts between states.

Finally, it would be derelict of me not to stress that whatever success the workshop achieved was due in large part to the hospitality and consideration shown us by the government of Bermuda. We are most appreciative of the participation of Sir Edward Richards, the premier, and also the Honorable Earle Seaton. Dr. Seaton not only helped make the workshop possible, but also sat through our lengthy debates and on many occasions, with his penetrating understanding of the position of smaller developing states, kept us from following our own logic beyond the boundaries of the possible.

Giulio Pontecorvo
Columbia University
April 16, 1974

Welcome Address

Sir Edward Richards, Premier of Bermuda

I am very pleased to be here this morning to welcome to Bermuda the Workshop
of the Law of the Sea Institute. It is the first time that you are holding your
deliberations here and, with regard to your discussions, I can think of no venue
more appropriate than Bermuda.

I must congratulate the Law of the Sea Institute on its devotion to
its aims and ideals, and on the reputation which it has achieved in the pursuit of
these aims over the past years. I am aware that it comprises many universities
and other distinguished institutions, and its study and research in oceanography
have attracted scholars of the highest repute. It is therefore not surprising that
its views have been of the highest importance to policymakers both in government and in the private sector—and would continue to be so.

Today a laissez faire attitude towards the problems of the sea is no
longer possible. For one thing, resources of the land are becoming scarcer and
men have to look to the sea for its contribution to food, oil, and other minerals.
Because the land mass of the Earth is less than one-third of the Earth's surface,
the oceans and seas present themselves as inexhaustible sources of supply for the
needs of mankind. The present energy crisis has served to intensify interest in
the unexplored resources of the seabed.

Since the end of the Second World War, attempts have been made by
many states to exercise sovereign rights over areas of the sea. By the Truman
Proclamation of 1945 the United States asserted sovereign rights over the
mineral and fisheries resources of its continental shelf. This was followed a few
years later by Chile, Ecuador, and Peru, who claimed jurisdiction over the ocean

200 miles off their respective coasts, and then began to seize U.S. fishing fleets found within those limits and to prosecute their captains.

To try to solve these problems of territorial limits and jurisdiction over fisheries as well as continental shelves, the United Nations held two conferences, one in 1958, the other in 1960. Success here was extremely limited. Since then the cod war between Britain and Iceland has flared up; an uneasy calm now prevails. It is therefore becoming imperative that uniform laws regulating these matters receive international agreement. To this end the United Nations will be holding another conference this year in Caracas, and I am sure that the views of the Law of the Sea Institute will make a fundamental contribution towards any success that may be achieved.

Situated as we are, the law of the sea is of vital interest to us in Bermuda. Our boundaries and the extent of our jurisdiction over resources for the purpose of policing, customs, and sanitation depend upon the definition of our territorial limits. About three or four years ago we made strenuous but unsuccessful attempts to have our territorial waters for fishing extended from 12 miles to 25 miles so as to include Challenger and Argus Banks; this was done because of reports that these fishing grounds were being depleted by foreign fishing fleets, fishing there at night. We will therefore be awaiting with considerable interest the results of your deliberations, for they may be of enormous benefit to the future of our local commercial fishermen.

We in Bermuda are not a party to any fishing dispute at present and, therefore, we are able to provide you with a serene and impartial venue. We hope that you and your spouses will be able to put Bermuda's reputation for hospitality and friendliness to the test during and at the end of your workshop. We are pleased to note that one of our own distinguished sons, Dr. Earle E. Seaton, our Puisne Judge, will be a participant in your workshop. He has had considerable experience in conferences relating to the law of the sea—and also in twisting the arm of government to host workshops. It is primarily through his initiative that we have the pleasure of hosting your workshop at this time. Associated with Dr. Seaton are the curator of our aquarium, Mr. James Burnett-Herkes, our fisheries expert, and Dr. John Arnell.

I wish you all every success.

Fisheries Conflicts in the North Atlantic: Problems of Management and Jurisdiction

Chapter One

Geography of the North Atlantic

Lewis M. Alexander

The concept of ocean basins as geographic regions carries with it certain definitional questions. Where are the limits to the region? What are its subdivisions? What features characterize it as a separate entity? How does it compare with other regions of similar magnitude? Are there unifying elements within the region as defined?

While experts and laymen alike often refer to the "North Atlantic" as a geographic unit, there is in fact no generally recognized definition as to just what is, and is not, included in the term. Presumably the North Atlantic lies north of the Equator, as distinct from the "South Atlantic," and somewhere at its northern extremeties it abuts on the Arctic Ocean. Does "North Atlantic" also include the North and Baltic Seas, the Mediterranean, and the Gulf of Mexico/Caribbean Basin? Is the Gulf of St. Lawrence part of the North Atlantic system, or of Canada's inland waters? This paper is not intended to settle, once and for all, such definitional questions? it seeks merely to highlight certain geographic issues which may be germane to the workshop we are holding here.

Although the Equator has traditionally been taken as the dividing point between the North and the South Atlantic, I would suggest that a more meaningful line would be from Cape Sao Roque, Brazil, to Cape Palmas, Liberia—that is, from 5° south latitude on the west to 4½° north latitude on the east. At its northern margin the region would abut on the Arctic Ocean at Robeson Channel between Canada and Greenland. East of Greenland, the division would approximate the 71° north latitude line joining Carlsberg

Fjord, Greenland, with Jan-Mayen Island and North Cape, Norway. This includes the Greenland and Norwegian Seas within the Atlantic region, leaving the Barents Sea in the Arctic.

Together with these northern and southern limits, another dividing line is important in the North Atlantic area—the one separating what might be termed the "northern" from the "north central" Atlantic. Such a line would extend from the Straits of Florida on the west to the Strait of Gibraltar on the east. Like other boundaries in the region it trends from southwest to northeast, moving from 23° to 36° north latitude between its western and eastern extremities. This is a line separating the "developed" from the "developing" North Atlantic; the utilized from the less utilized, so far as fisheries are concerned; the well traveled from the lightly traveled, shipping-wise. The northern and north central areas might then be subdivided into eastern and western components, giving us Northeast, Northwest, East Central, and West Central sectors.[1]

There remains finally the question of inclusion or exclusion of marginal seas within the North Atlantic system. The answer, it seems to me, is a relative one. For some phenomena, such as shipping or military strategy, all marginal water bodies should be included; in other considerations—as for example, water movement or fisheries operations—one or more of the semi-enclosed seas may represent something of a subsystem, largely independent of the rest of the oceanic basin. The seas involved are the North and the Baltic, Baffin Bay/Davis Strait, the Gulf of St. Lawrence, the Gulf of Mexico/Caribbean Basin, and the Mediterranean/Black Sea.

PHYSICAL CHARACTERISTICS

The major axis of the North Atlantic trends for nearly 5,000 miles,[2] from the Caribbean to the Norwegian Sea. Most of the major physical features of the Basin are also oriented in this general direction. The principal climatic element affecting the movement of winds and of water is the semi-permanent high pressure system, centered at about 35° north latitude throughout the year, although becoming much stronger in summer than in winter. About this system the winds move clockwise, with the northeast trades to the south moving onto the coast of South America, and the prevailing westerlies to the north bringing air masses from North America to Europe.

Moving clockwise about the high pressure mass is the Gulf Stream system fed by the North Equatorial Current which flows east-to-west from North Africa to the West Indies. One branch of the current passes through the Caribbean and Gulf of Mexico, emerging through the Straits of Florida as the Gulf Stream. It is deflected to the northeast by the Great Bahama Bank, and flows roughly parallel to the U.S. coast to the vicinity of Cape Hatteras; here it veers more to the east, passing close to the Grand Banks. The current then turns east and later subdivides, one branch moving southeast and south as

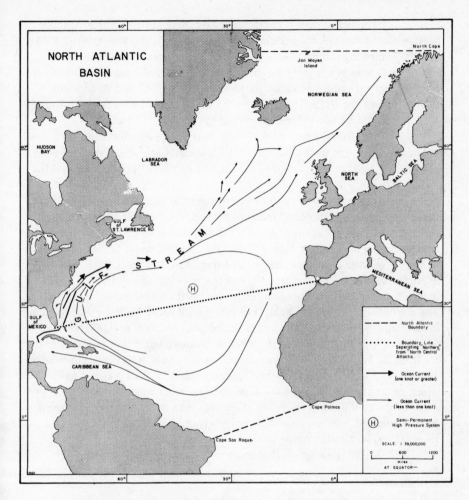

Figure 1-1. North Atlantic Basin

the cool Canaries Current, while the other, the warm North Atlantic Drift, flows to the northeast off the British Isles and into the North and Norwegian Seas. The other major current of the North Atlantic is the cold Labrador Current, which moves down through Davis Strait and the Labrador Sea to the vicinity of the Grand Banks.

Most of the North Atlantic is ice-free in winter. The mean January isotherm of 32°F. extends from Cape Cod to southern Newfoundland, southern Iceland, and into the Barents Sea. Only to the north and west of this general line does shipping in wintertime tend to be seriously impeded by its conditions.

The ocean floor is characterized by three major features. Curving north-to-south through the middle of the ocean basin is the Mid-Atlantic Ridge, a 10,000-mile-long submerged mountain chain which reaches the surface to form islands only in scattered patterns. The Azores are the principal island formation of the mountain chain; far to the north, Iceland and its adjacent shelf sit astride the Ridge. Splitting the Ridge at its crestline is the Mid-Atlantic Rift, a fissure some two miles deep and up to thirty miles in width.

The Ridge separates the North American, Labrador, and other oceanic basins to the west from the West European, Canary, and similar basins on the east. Smaller basins exist in the Caribbean, the Gulf of Mexico, and the Mediterranean.

From a resource point of view, the most significant features of the ocean floor are the continental margins, particularly the shelf areas. Under the North Atlantic proper are approximately 1 million square miles of shelf; another half million square miles underlie the marginal water bodies—the Gulf of Mexico and the Caribbean, Baltic, Mediterranean, and Black Seas. The importance of these shallow areas to fisheries resources is well known. And as mining interests begin to focus on the deeper portions of the continental margin, new configurations come to light. For example, Rockall Bank, west of Scotland, has a relatively small shelf adjacent to it at depths of 2,000 to 6,000 feet.

One final aspect of the physical environment is that of access to the sea from the land. The North Atlantic area is fortunate in its many indented coastlines, its bays and estuaries and offshore islands, and its seas and gulfs which penetrate deep into the mainland. In few other areas of the world is so much of the land within easy reach of the sea. And in few other ocean basins is travel possible between or within continents using such relatively short distances from one landfall to the next.

All of these physical characteristics, taken together, make the North Atlantic a very "usable" ocean body.

COMMERCIAL FISHERIES ACTIVITIES

Since this workshop is focused largely on fisheries issues in the North Atlantic, I shall treat the subject matter here only lightly. Gulland[3] has estimated the total fisheries potential in the North Atlantic to be in the range of 25 to 29 million tons live weight.[4] These figures break down roughly as follows:

	Low Estimate (000 tons)	*High Estimate (000 tons)*
Northwest Atlantic	7.00	7.00
Northeast Atlantic	10.30	11.80
West Central Atlantic	2.00	2.70
East Central Atlantic	2.10	2.70
Gulf of Mexico	1.70	2.40
Caribbean	.45	1.00
Mediterranean/Black Sea	1.50	1.50

The 25–29 million ton figure for the North Atlantic region compares with the following general estimate of fisheries potentials in other ocean areas (in millions of tons):   South Atlantic, 12-13;   North Pacific, 11;   Central Pacific, 16-24;   South Pacific, 14;   Indian Ocean, 14.   These figures, taken from Gulland, are extremely rough, and are intended only to indicate relative orders of magnitude.   But they do point out that the Atlantic, north of the Equator, is one of the world's foremost areas in terms of estimated fisheries potential.

BORDERING STATES AND THEIR OCEAN INTERESTS

The countries bordering on the North Atlantic might be divided into the following groups: (1) Anglo-American; (2) northern Latin America, including the Caribbean countries and those of northeastern South America; (3) Atlantic Europe; (4) northwestern Africa; and (5) the Mediterranean/Black Sea. Altogether, 61 states are represented here, together with a number of overseas dependencies of colonial powers. Taken together, these states' populations comprise 1.325 billion or about 40 percent of the world total.[5]

In the above groups, all of the states of Anglo-America and Atlantic Europe would probably be classed as "developed," while none of the countries of the Latin American or African ones would. The Mediterranean/Black Sea states would be divided, those of Europe (including Malta and Cyprus) are presumably in the "developed" category, and those of southeastern and southern Mediterranean (with the exception of Israel) are not—although Libya, with its oil wealth and its small population, is something of a borderline case. There are other borderline cases as well, among them Venezuela with its oil and iron ore, Trinidad and Tobago with its oil, and in time, perhaps, Jamaica and Cuba. New independent countries will emerge, the first of these being Granada, which is scheduled for independence from Britain in February 1974. Belize, the former British Honduras, will achieve self-rule soon, and some of the other Carribean dependencies may also become free of foreign control in the near future. Unfortunately both Belize and the Caribbean islands would appear at this time to lack the natural resource base necessary for a viable economy as independent units.

NATIONAL OCEAN INTERESTS

The ocean interests of coastal states might be grouped under the following headings: commercial fishing, offshore mining, shipping, recreation, industrial uses, scientific research, and military activities. It is not the intent of this paper to describe in detail the seven categories of marine interests for the 61 states and the various dependent areas of the North Atlantic region. The text, rather, will be limited to a few remarks concerning each group of uses.

Of the ten major fishing countries of the world, six border on the

North Atlantic, and carry out a major share of their fishing activities there.[6]
The economic dependence of Iceland on fishing has long been noted, a situation
which also prevails in the Faroes, and in regional sectors of other countries such
as northern Norway, eastern Canada, northwestern France, northwest Spain,
and—to a declining extent—eastern New England. Even if national economies
are not strongly oriented toward fisheries, sectional interests may be, and these
interests are sometimes able to pressure governments into taking strong measures
to protect their local industry against foreign competition.

Offshore mining has to date been confined to oil and gas operations,
although the era of hard mineral exploitation is about to begin. Three areas of
oil and gas activites predominate in the North Atlantic—the U.S. Gulf Coast,
Venezuela's Lake Maracaibo, and the North Sea. But explorations are being
carried out now in widely scattered sections of the Atlantic region. Soon, new
areas may be in production, among them northeast United States and eastern
Canada and the waters off eastern Spain. There are also indications of significant manganese nodule concentrations, particularly in the western North
Atlantic, including the Blala Plateau off southeastern United States.

Shipping, a third category of marine interest, has long been a major
activity, particularly in the northern Atlantic. East-west traffic between Anglo-
America and Atlantic Europe is the heaviest of any transoceanic movement in
the world, but considerable shipping also takes place from northern Latin
America to both the United States and Europe. Although the closure of the
Suez Canal has affected the Mediterranean's transit role, there still is a high
volume of movement of ships between that area and both Western Europe
and the United States. Coastwise shipping is also of great importance to the
countries of the northern Atlantic, contributing to the growing problems of
pollution control and of safety at sea. Of the leading countries of the world in
merchant marine tonnage, all but Japan are located on the North Atlantic.

A growing issue with respect to shipping is that of the movement of
oil and natural gas. On the one hand there is the question of port facilities, as
more and more super tankers are put into service. Where will superports and
new refineries be located? Will new patterns of oil production and movement
come into being as a result of real or potential Arab oil embargoes? What new
safety regulations may be put into effect for tankers and LNG carriers? Western
Europe and the United States are, along with Japan, the major world importers
of waterborne petroleum. Among their paramount interests is ensuring that
sources of supply continue to be available, even perhaps at the expense of
downgrading their commitments to other aspects of marine use.

Recreation is thought of here as of two types—the enjoyment by
a nation's citizens of the recreational attributes of the marine environment, and
the development of these attributes commercially as a source of income from
foreign visitors. Although in both cases the recreationists and their spokesmen
would be interested in preserving the quality of the marine environment to the

greatest extent possible, in the latter situation government leaders might be extremely sensitive to the dangers of pollution of coastal areas, possibly to the extent of closing off straits and other water areas to use by tankers and other potential polluters. Such conditions may well come to pass in island groups in the Caribbean, in the Bahamas, and in the northern and northeastern Mediterranean.

Industrial use covers two types of operations: the utilization of water for cooling, desalination, or other commercial purposes; and the use of the sea as a disposal area for wastes. The second category is obviously the more important in terms of multi-use conflicts. The affluent nations of the northern Atlantic have long been accustomed to regarding the sea as a convenient disposal area for unwanted substances. Now, with the growing concern for environmental protection and preservation, all sorts of new questions arise. How will permissible levels of disposal be determined? How can regulations be implemented against foreign vessels? What policies will the nondeveloped countries adopt in terms of pollution control and abatement? What efforts should be made not only to halt degradation, but to start cleaning up polluted areas of the marine environment? In semienclosed seas pollution problems may become particularly severe. Here is perhaps one of the most divisive international issues associated with the use and control of the sea—an issue which is very pertinent to the North Atlantic, and which could pit some of the developed countries of the northern Atlantic against developing states of the North Central area.

Scientific research is also a potentially divisive element. On the one hand there are the far-reaching oceanographic activities of the United States, the U.S.S.R., the United Kingdom, France, and other developed states; there are, on the other hand, the suspicions and impatience of many developing countries, which are reluctant to grant permission for foreign research close to their coasts, even after receiving such guarantees as open publication of research results and a chance to participate in the oceanographic activities. The gap between rich and poor countries, so far as scientific capabilities are concerned, may be widening even faster than that of per capita GNP or other indices of wealth. Few viable mechanisms seem to exist at this time for helping to bridge the gap, either among North Atlantic countries or with respect to other areas as well.

Finally, there are military interests. Despite any seeming detentes, the two superpowers face one another across the North Atlantic and in the Mediterranean, their military postures augmented by the presence of air and sea bases and of military allies occupying the geographic space between Washington and Moscow. Keflavik, Lages, Cienfuegos, Rota, Holy Lock—these and other names attest to the continued confrontation stance in the ocean and its borderlands, a stance that is made even more complex by the existence of other strategic concerns as well, such as those of Britain, France, and the Netherlands with their overseas territories, of Israel and the Arab states, and of newly emerging powers such as Brazil in establishing their presence in the waters off their own coast.

Two pictures emerge from this general discussion of national ocean interests in the North Atlantic. One, involving the ocean space, is that of a heavily utilized northern Atlantic, but with activities gradually moving south-ward both through increased resource use in the Gulf of Mexico/Caribbean Basin and off northeastern South America, and in terms of fisheries and tourism, off western Africa. Within the northern Atlantic problems of congestion are becoming extremely severe in such areas as the northeastern United States and the North Sea, particularly as the growing sea-oriented oil and gas activities are superimposed on the already complex use patterns. The need for new rules and regulations to handle multi-use conflicts is clearly evident, but resolution of these conflicts is proceeding at a glacierlike pace.

The other picture is that of two groups of countries bordering the North Atlantic with generally dissimilar ocean interests. The developed states of the northern Atlantic have traditionally been ocean-oriented; those of the North Central area have not. There are variations within these two groupings. The Soviet Union, and several of the East European socialist states, have been in-vesting far more heavily in marine-related activities over the past two decades than in former years. Israel, in less than three decades of independence, has built up a substantial marine industrial complex. Both Canada and Brazil have come to regard their adjacent ocean areas with far greater interest than in the past. But the basic dichotomy between ocean interests of the northern and southern countries of the North Atlantic persists, thus presenting an obstacle to regional cooperation in marine resources development.

THE NATURE OF OFFSHORE CLAIMS

Three aspects of offshore claims will be treated here: (1) the baselines from which zones of offshore control are measured; (2) breadths of the territorial sea; and (3) assertions of control beyond territorial limits.

The traditional method of establishing baselines along the coast is to follow the low-water line, making use of the exceptions provided for in the articles of the 1958 Geneva Convention on the Territorial Sea and Contiguous Zone. The Convention also permits the use of special straight baseline regimes under certain circumstances. Of the 60 coastal states in the North Atlantic area, 31 had by 1973 either established straight baseline regimes along all or a part of their coasts, or enacted legislation permitting such an establishment.[7] In a few cases, historic bays or other water areas were closed off as internal. Among these were Poland's Bay of Gdansk, Portugal's Tagus and Sado Estuaries, Guatemala's Bay of Amatique, and Egypt's Bay of El Arab. The Soviets claim the Bay of Riga and the Sea of Azov as historic waters.

To date, none of the island groups of the North Atlantic have been joined together under the "archipelago principle" except for the Faroe Islands, where a closing line about the group serves as a basepoint for an exclusive fisheries zone.

Turning to the breadth of the offshore zones, most states have adhered to the principle of 12 miles as a maximum limit. Either their territorial sea extends to this distance, or, if the breadth is less, there generally is a contiguous zone (which may or may not include exclusive fisheries provisions) between the outer territorial limits and 12 miles from shore. Gambia claims a 50-mile territorial sea, Guinea's extends to 130 miles, while 200 miles is claimed as the breadth of the territorial sea by Brazil, Panama, and Sierra Leone.[8]

Beyond the 12-mile limit there are several fisheries claims. Haiti's exclusive fishing zone extends to 15 miles offshore, Iceland's to 50 miles, Morocco's to 70, Senegal's to 110, and Nicaragua's to 200. France, in 1972, limited foreign fishing to 80 miles off the coast of French Guiana. With respect to these fisheries zones, care must be taken to distinguish among various claims to competence (i.e., exclusive fishing, conservation rights, etc.). In some cases the legislation is unclear as to what restrictions on foreign activity within the fishing zone are actually involved.

Five states—Dominican Republic, Haiti, Egypt, Syria, and Venezuela— have claims relating to such functions as customs, security, and sanitation out to 15 or 18 miles off their coasts. Canada, in 1970, extended pollution control regulations out to 100 miles from land north of 60° north latitude. In the same year it established fisheries closing lines in the Bay of Fundy, Gulf of St. Lawrence, and Queen Charlotte regions. The following year its pollution control measures were extended to all Canadian waters south of 60° north. And since 1939 the United States has claimed a 300-mile defense zone around the entire hemisphere, excluding Canada.

It is difficult to detect a pattern to the offshore claims in excess of 12 miles. The 200-mile bloc of South American states has not yet succeeded in converting most of the Caribbean countries to its point of view, and the transfer of 200-mile claims to Africa has affected only Sierra Leone. Perhaps during the deliberations of the Third Law of the Sea Conference the pattern of territorial and fisheries claims beyond the 12-mile limit may begin to change.

Two other points with regard to offshore claims in the North Atlantic are worthy of note. A first concerns the delimitation of seabed boundaries between opposite or adjacent states. With the increase in interest in offshore oil exploration and exploitation, the issue of seabed boundary delimitation becomes a very real one. Most of these boundaries in the North Sea have been delimited, the two exceptions being those of Belgium and France. In the Baltic the Soviet Union has concluded seabed boundary agreements with Poland and Finland, and Poland with East Germany. Italy's boundary with

Yugoslavia in the Adriatic has been settled, as has the U.S.-Mexican boundary to 12 miles from shore. Other than these, however, no formal agreements on the delimitation of seabed boundaries in the North Atlantic area have been concluded.[9] With continued leasing and exploratory operations taking place, some of these unresolved issues may become quite acute, as for example, in the Gulf of Maine where both Canada and the United States are actively interested in offshore oil and gas developments.

A second problem concerns the right of innocent passage through international straits—for military vessels and aircraft, for potential polluters such as oil tankers and LNG carriers, and for other forms of shipping. The Strait of Gibraltar is perhaps the site of the most serious confrontation potential between coastal and user states, but some of the straits connecting the Caribbean with the Atlantic, those passing through the Bahamas and those in the northeastern Mediterranean, may also in time be involved in questions of freedom of transit. Here again is a potentially divisive force among the countries of the North Atlantic area.

THE NORTH ATLANTIC AS A GEOGRAPHIC REGION

At the start of this paper several questions were raised concerning the regional concept as it applies to ocean basins. Two in particular seem relevant to any summary analysis. What are the elements of unity and diversity in the region, and how does this one compare with other oceanic regions of the world?

It is not difficult to enumerate many forces for unity and for diversity within the northern Atlantic. The press is continually reviewing relations within and between Anglo-America and Atlantic Europe; to a considerable extent these general forces are reflected in commonalities and divergences of marine-related interests and policies on the part of these states. The littoral countries of the Black Sea have many interests in common, although this situation is less evident in the Mediterranean Basin. Indeed, centripetal forces, particularly between the northern and southern rims of the sea, have tended to be on the increase in the years since World War II.

There is little commonality of interests to date between northern Latin America and northwestern Africa. Trade between the two areas is minuscule, and there are few ties of history or common cause to bring the two areas together. Northern Latin America, on the other hand, has strong economic, political, and cultural bonds both with Anglo-America and Atlantic Europe, while Northwest Europe in turn has ties (both of positive and negative types) with Northwest Africa. Anglo-America, on the other hand, has few such ties.

What does all this suggest? Simply that, in the interests of regional development of the North Atlantic marine environment and its resources, the countries of Anglo-America and Atlantic Europe might place higher priorities than they have in the past in seeking to cement more firmly their ties with the North Central Atlantic states on both sides of the ocean. This should be especially true with respect to the marine-related activities of these countries. While it is true that the United States, Canada, and the U.S.S.R. also have interests and commitments in the North Pacific and Arctic Oceans, this should not detract from their concern with management and development of the North Atlantic environment. In this area the rich countries are in close proximity to poor ones, and the ocean space provides a geographic link between them.

Comparing one ocean basin with another is, at best, a risky business, since the concept of ocean regions itself is still in an evolutionary form. What other regions exist as geographic units? Which of their characteristics are capable of comparison? The South Atlantic and Indian Oceans are still decades away from the type of development the North Atlantic has experienced. The Arctic is bordered by areas that are largely empty of settlement. The Pacific, in whatever sectors it may be divided, is probably the most viable candidate for comparison with the North Atlantic. I would suggest a general division of that ocean into three parts; the North, the Southeast, and the Southwest.

The North Pacific, like the North Atlantic, is a region of considerable oceanic use, including shipping, fishing, and scientific and military activities. Unlike the North Atlantic it is not bordered by heavily populated areas, except along its western rim. I would use the Tropic of Cancer (which bisects Taiwan) as the southern limit of the North Pacific. North of this line most of the development is in marginal areas; the East China/Yellow Seas and the other seas to the northeast on the Asian rim; and on the east the North American coast from Baja California to Alaska.

The Southeast Pacific is a vast expanse of water extending from Samoa eastward to South America. It has few islands or shallow areas, and few known fisheries resources. West of the date line, however, is the Southwest Pacific, stretching past Australasia, Indonesia, and the Philippines to the Asian mainland. This area seems to me to be one of enormous potential. Although, like the South Atlantic and Indian Oceans, it is still many years away from extensive development, it has the potential in time of perhaps exceeding the North Atlantic in wealth and in importance to a substantial segment of the world population.

The lesson to be learned from any attempt at comparing the North Atlantic with other regions is this. While the North Atlantic continues today as a leader in ocean use and management, it faces continuing problems of congestion, pollution, and resource depletion, which in many respects are unique to this part of the world. Other oceans are behind in the processes of both development and deterioration.

Notes to Chapter 1

1. These subdivisions are utilized by FAO in their publication, *Fish Resources of the Ocean*, compiled and edited by J. A. Gulland (Rome: Food and Agricultural Organization of the United Nations, 1970), although the boundaries of the four regions differ somewhat from those suggested here. FAO's statistics for the Northeast and Northwest are based in part on ICNAF and ICES data; therefore, the regions conform with the limits set by these two organizations.

2. All mileage figures are in nautical miles. One nautical mile equals 1.15 statute miles. One square nautical mile equals 1.324 square statute miles.

3. See note 1.

4. Estimates for the ICES (International Council for the Exploration of the Sea) area include the Barents Sea and the Spitzbergen/Bear Island waters which are outside the limits, as defined here, of the North Atlantic. An estimated 10–15 percent of the total potential, as listed in *The Fish Resources of the Ocean*, may be in these Arctic waters.

5. The states included in each group are as follows. Some states obviously fall into more than one group. (Major dependencies are underlined.)

Anglo-America (3): U.S., Canada, Bahamas, Greenland (Den.)

Northern Latin America (18): Mexico, Guatemala, Honduras, El Salvador, Nicaragua, Costa Rica, Panama, Colombia, Venezuela, Cuba, Haiti, Dominican Republic, Jamaica, Granada, Trinidad and Tobago, Barbados, Guyana, Brazil, Belize (U.K.), Puerto Rico (U.S.), Surinam (Neth.), French Guiana (Fr.), Caribbean Islands (U.K., Neth., Fr.).

Atlantic Europe (16): Iceland, Ireland, U.K., Norway, Sweden, Denmark, Finland, U.S.S.R., Poland, East Germany, West Germany, Netherlands, Belgium, France, Spain, Portugal, Faroes (Den.).

Northwestern Africa (7): Morocco, Mauritania, Senegal, Gambia, Guinea, Sierra Leone, Liberia, Spanish Sahara (Sp.), Guinea-Bissau (Port.) Azores (Port.).

Mediterranean/Black Sea (17): Monaco, Italy, Malta, Yugoslavia, Albania, Greece, Turkey, Cyprus, Syria, Lebanon, Israel, Egypt, Libya, Tunisia, Algeria, Bulgaria, Romania, Gibraltar (U.K.).

6. The United States, for example, derives about 70 percent of its catch, by volume, from the Atlantic Ocean and the Caribbean. The four non-Atlantic fishing states among the top ten in 1972 were Peru, Japan, People's Republic of China, and Thailand.

7. These states were: Albania, Brazil, Canada, Cuba, Denmark, Dominican Republic, Egypt, Finland, France, Gambia, East Germany, West Germany, Guatemala, Guinea, Haiti, Iceland, Ireland, Malta, Mexico, Norway, Panama, Poland, Portugal, Senegal, Sweden, Syria, Turkey, U.S.S.R., U.K., Venezuela, and Yugoslavia. Data from *International Boundary Study, Series A,*

Limits in the Seas, "National Claims to Maritime Jurisdictions," No. 36, March 1, 1973 (Washington: U.S. Department of State, Office of The Geographer, 1973).

8. The 200-mile territorial claim of Honduras is still on the books, although the 1965 Constitution accepts a 12-mile limit.

9. An agreement has been reached on the delimitation of the shelf boundary between Canada and Greenland in Baffin Bay/Davis Strait, but has not yet been ratified.

Chapter Two

North Atlantic Fishing Adaptations: Origins and Directions

Raoul Andersen

INTRODUCTION

This paper explores the background or evolution of approaches to fish resources among peoples of the North Atlantic rim in an endeavor to illuminate the present direction and status of fisheries in this zone.[1] In doing so, though drawing in part upon archaeological sources, my discussion is largely limited to a cultural-social anthropological summary of what appear to be fundamental trends affecting the maritime-fisherman peoples and fisheries of the North Atlantic. The relationship between these trends and the issue of fishery juris-diction and enforcement is also explored, as are some implications for fishing and fisherman occupations and communities perhaps inherent in the new regimes expected from the Law of the Sea enterprise.

THE DEVELOPMENT (EVOLUTION) OF NORTH ATLANTIC FISHING

At the most general level, admitting the great range and diversity among fishery environments, human histories, and cultures found in and around the North Atlantic littoral, we may distinguish several analytically distinctive trends operative in fishery cultural development in this area: (1) Increasingly effective technologies characterized by mobility, versatility, and capital intensity; (2) an expanding (spatially and in intensity) use of natural resources; (3) increasing occupational specialization apparent in the shift from subsistence to labor production; and (4) the growth of increasingly complex forms of integration linking fishing production with local, regional, state, and international markets, economies, and societies.

These trends are simultaneously expressions of the evolution of culture in general. They underly the emergence and present state of both the "global society," and, more concretely, Law of the Sea fishery resource management goals.

These trends are only analytically separable from each other in the same sense that, speaking concretely, the twentieth-century development of sophisticated electronic fish-finding and navigational technology centralized in the wheelhouse of a modern fishing vessel makes hunting and catching more efficient (as regards time and energy investment, not to mention financial and other forms of capital investment), and, at the same time, extends exploitative control over marine resources and relies upon and contributes to the importance of formal technological education (e.g., fisheries colleges) in order to be fully utilized. This increases the fisherman's integration with and dependence upon the complex techno-scientific institutions and knowledge of his society. For example, where the skipper formerly might simply acquire his skills from his father or another fisherman on the job, he is now pressured to "institutionalize" himself for varying periods to acquire requisite skills and know-how. The oft-mentioned notion of having "professional" fishermen is a concrete manifestation of this trend towards increasing specialization and complex forms of integration. It also expresses increasing concern over problems in recruitment, generated in part by the socioeconomic implications of new occupational regimes structured around these new technologies in labor markets with many alternatives to careers in fishing.

The complex system of dynamic interrelationships in which modern commercial fishermen of North Atlantic and other fishing nations operate is only partly expressed above. Viewed against the long path of human evolution, this dynamic and complex order is remarkable for its relative recency, rapid change, and the great difficulty it poses in our attempts to mold it to human aspirations. What follows expresses an anthropologists' bent to see this rapidly changing order as manifesting the aforementioned general cultural evolutionary trends which give rise to increasingly complex production systems.

FROM PREHISTORIC TO HISTORIC
NORTH ATLANTIC FISHERIES[2]

The evolution of fish-resource exploitation is usefully explored for present purposes by examining the general succession of ecological adjustments or "ecotypes" that developed in advance of our modern North Atlantic industrial fisheries. At some risk of oversimplification, the following discussion sketches these ecotypes, viz., hunters and gatherers, fishermen-farmers, and fisherman specialists.

Hunter-gatherers

It is difficult to say how long marine fauna and flora have been included in the total spectrum of man's ecological adjustment. But there is strong opinion among prehistorians that the resources of river and sea were in significant use only late in the evolution of man as hunter and gatherer. Washburn and Lancaster summarize the situation as follows:

> During most of human history, water must have been a major physical and psychological barrier and the inability to cope with water is shown in the archaeological record by the absence of remains of fish, shellfish, or any object that required going deeply into water or using boats. There is no evidence that the resources of river and sea were utilized until this late preagricultural period, [ca. 20,000–30,000 years ago] and since the consumption of shellfish in particular leaves huge middens, the negative evidence is impressive. It is likely that the basic problem in utilization of resources from sea or river was that man cannot swim naturally but to do so must learn a difficult skill. In monkeys the normal quadrupedal running motions serve to keep them afloat and moving quite rapidly.
>
> For early man, water was a barrier and a danger, not a resource. (Obviously water was important for drinking, for richer vegetation along rivers and lakeshores, and for concentrating animal life. Here we are referring to water as a barrier prior to swimming and boats, and we stress that, judging from the behavior of contemporary apes, even a small stream may be a major barrier.)[3]

With respect to the peoples of the North Atlantic rim, our focus in this paper, prehistorians credit the human record with evidence of many successful local maritime adjustments in the early postglacial period or late in the Upper Paleolithic roughly between 22,000 and 7,000 years B.P. (before present). In Europe, in a warmer, more moist climate, open vegetation was gradually replaced with dense forests and led to new species of game (e.g., reindeer, horse, and bison gave way to small deer, elk, and pigs), both of which made land hunting more difficult and caused the forest inhabitants to settle along salt- and fresh-water margins where fish became a staple food.[4]

Numerous fish remains have been found associated with Magdalenian cultural material dating between 22,000 and 11,000 B.P. in sites along the Dordogne and Vezere river valleys of southern France.[5] These river and seaside peoples developed a great variety of nets, hooks, and other fishing gear such as fish traps, and dugout canoes and paddles, all adding to the "equipment

of the chase" for fish in relatively shallow waters of lakes, rivers, and the inshore coastline. But these remained essentially pluralistic land *and* coastal-
riverine hunting and gathering adjustments throughout the Mesolithic (ca.
12,000 to 8,00 B.P.) and well into the Neolithic agricultural period. Even
where the archaeological record suggests relatively sedentary—as opposed to
seasonally based—populations established at water's edge, e.g., the Ertebølle
people who left great shell and bone middens along the shores of Denmark,
both land and marine food and material resources were being exploited.

The development of maritime adjustments on the western North
Atlantic littoral follows a different absolute chronology, and is later in evidence
of coastal-estuarine marine resource use. But the North American counterparts
of essentially Upper Paleolithio–Mesolithic European cultures were using east
coast fresh- and salt-water resources at least by 4000 B.P.,[6] and perhaps earlier
than 10,000 B.P.[7] These were similarly hunter-gatherer peoples who relied
upon both land *and* water resources. Well before historic European contact,
but presumably independent of their European counterparts, they too had
developed a complex body of fishing techniques (e.g., weirs, traps, seine and
gill nets, spears, harpoons, hooks, and canoes) which enabled them to exploit
a broad shallow-water or inshore marine resource spectrum.[8]

The Mesolithic (to about 4500 B.P.) semi-nomadic sub-Arctic
hunting-fishing regime found in North Norway, while not generalizable in its
particulars (especially species cited) to other areas, is suggestive of the broad,
seasonally varied ecological base that developed around the North Atlantic rim:

> In the summer time their economy was based primarily on fishing,
> sealing and whaling, bird catching and gathering of eggs and down,
> in some periods even on the gathering of snails and shells. In the
> winter season hunting with reindeer, elk and bear as leading victims,
> and fresh-water fishing through the ice were important pursuits.[9]

Whether we speak of the nomadic/semi-nomadic hunter-gatherer
groups, which exploited fresh- and/or salt-water fish, or of relatively sedentary
coastal populations largely based upon intensive fishing,[10] these Mesolithic
peoples were organized into essentially small, undifferentiated, autonomous,
and subsistence-oriented groups. It is further reasoned that among these small
groups, rights in basic subsistence means—hunting, fishing, and gathering territories—if defined at all, pertained to the band in common. Such group ownership normally also embraced the bulkier gear used in the food quest, such as
boats or weirs.[11]

There was local environmental specialization in these adjustments,
but there is no evidence of an extensive trade involving fish between coastal-
riverine and other inland hunters and gatherers. Doubtless, some secondary
products, especially shell and bone, derived from marine resource extraction

were traded (worked and unworked) in many areas. This perhaps increased local extractive efforts somewhat beyond immediate subsistence requirements in order to obtain locally unavailable material resources, e.g., scarce flint, or superior products, from other nonagricultural areas.[12]

Archaeological evidence also reveals that, on the technological level, the increase in the intensity and diversity of marine resources exploited by North Atlantic littoral peoples from the Upper Paleolithic to early Neolithic era is paralleled by development of increasingly efficient tools, e.g., spears, nets, lines, and hooks fitted to specialized catching tasks. The development also involved advances in the kinds of materials used for tools and in their manufacture. For example, in coastal northeastern North America and Scandinavia, less efficient (wasteful of labor and material resources) microlithic *chipped* blade tools were gradually displaced by *ground* slate tools which could be easily fashioned to fit many special functions.[13] Similarly, Mesolithic barbless stone hooks gave way to barbed stone hooks, then to highly durable barbed metal hooks in the Neolithic cultures of Denmark and South Sweden.[14] Further, the development of dugouts and skin boats increased opportunities to use existing fishing gear, by changing rivers and seas from barriers to pathways to new resources and new areas for settlement[15]—a harbinger of the international competition for marine space and resources which concerns us here.

Fishermen-farmers[16]

Broadly based ecological adjustments involving hunting (including fishing) and gathering were only very slowly transformed by the advent of agriculture in the Old World. Agricultural knowledge had still to evolve a long way before wild food production would decline. Initially, fishermen-gatherers gradually included agriculture—inclusive of stock domestication (e.g., goats, sheep, horses)—in a broadened pluralistic seasonal and/or opportunistic subsistence (and, much later, market-oriented) adjustment.[17] This basic adjustment, now linked with industrial market economies, persists importantly in various patterns of the littoral, e.g., Tidewater Virginia, North Norway, Faroes, Nova Scotia, Newfoundland, Scotland (crofter-fishermen), and Iceland today.

One is at risk in attempting to specify a time or a place for its first appearance. It had surfaced, declined and vanished, and reappeared many times in various European coastal and riverine settings by 3000 B.P. It doubtless first developed in an eastern Mediterranean setting close on the heels of Neolithic agricultural development (ca. 10,000 to 12,000 B.P.), and later surfaced at various points around the eastern North Atlantic littoral where various combinations of Neolithic technologies and agricultural skills enabled men to extend control over land flora and fauna, and domesticate these resources.

In the New World, agricultural (though plowless, and hence *horticultural*) knowledge developed somewhat later (ca. 7,000 B.P.) in the areas of Mexico and Central America, from whence it spread through the Gulf coastal area, up the Mississippi and Missouri Rivers and their many tributaries, and by 1000 B.P. continued east to embrace coastal populations from upper Florida to Maine and New Brunswick. An adjustment involving maize, bean, and squash horticulture, and fishing for food and crop fertilizer, among other wild resource production, it was well established among the Amerinds when the first permanent European settlers arrived in the New England and Middle Atlantic or Tidewater Virginia areas. The Powhatan Indian horticulturalists and sturgeon fishers of the latter marine zone reported in the early seventeenth century are illustrative of this development.[18]

Fisherman specialists

The establishment of trade relationships between those European coastal populations advantageously situated to exploit marine resources, and, first regional, then more distant agricultural populations capable of conducting long-distance trade activities, greatly altered man's approach to fishing and the sea. Locally unavailable goods become a major and enduring incentive to produce fish beyond subsistence requirements. Human efforts to improve extractive and preservation technology and the organization of production on the marine and other fronts were intensified. This stress simultaneously provided conditions essential to development of an increasingly specialized community, then occupational subcultural specialization and, although there is no precise time when we may first use the unqualified term, *fisherman* labor production in various areas, certainly by the nineteenth century.[19]

Populations expanded under the influence of crop and stock domestication on both sides of the North Atlantic before European trans-Atlantic exploration and colonization made its quantum leap in the late fifteenth century. In Europe, freely available arable coastal lands grew scarce under increased demand, and there gradually emerged the new ecotype of rather specialized farmers removed from the coast and river valleys, yet linked with coastal or riverine fishermen-farmers with whom they bartered various surplus agricultural, forest, and other products in exchange for fish. There followed closely, if not simultaneously, the development of specialized communities of fishermen—many landless peasants—in coastal areas with lands marginal, if not worthless agriculturally, somewhat removed from the trade and population centers that emerged in conjunction with a growing agricultural and craft production and trade. Other specialist fishermen "communities" appeared, for example, in Baltic coast towns, as distinctive occupations within these trade centers.[20] Both communities became increasingly specialized primary producers for the burgeoning markets and urban populations that grew explosively with the Industrial Revolution.

Specialized fishing operations involving seasonal and/or permanent stations, or plantations closely linked with newly emergent trading corporations, e.g., fishing merchant guilds, had been developed by Phoenicians to supply

Levant markets and operated on the Atlantic coast of the Iberian Peninsula well before the Christian era. But, according to Coull,[21] it was not until the sixth or seventh centuries that the fisheries of some parts of the European Atlantic coast developed a fish trade; "and in the seventh century on the Friesian coast there were communities whose main activity seems to have been fishing." These are thought to have been *inshore* fisheries. It was not until the Vikings and Normans that the first "deep-water" fisheries were undertaken.[22] Coull writes:

> In the tenth century the Normans were pursuing deep-water fisheries, some of them probably with long-lines, from seasonal bases in the south of Ireland, and in the following century English fishermen were venturing into deeper waters for herring. The early commercial fisheries from the west coasts of Europe were principally for cod and herring, which are still the main species sought by the modern fleets.[23]

FISHERMEN PLURALISTS AND SPECIALIZATION

Coull's remarks above provide no details on the fisherman's total work regime or the general organization of Norman and English fisheries of this period. The fishermen in question were probably only seasonally occupied, and returned to small farms following cessation of fishing (and any subsequent trading outside their locale). We may further surmise that they worked as small crews of kinsmen who shared jointly in the problems of organizing and managing fishing capital, meeting risks, and sharing profits. Many doubtless also transported their finished products to distant markets where they bartered or sold directly.

Technologically, their putative "deep sea" fishing probably involved each fisherman working one or two handlines equipped with a few baited hooks and sinkers, from the rail of an open boat in waters of about 15 to 20 fathoms depth. The "bultow" or line trawl commonly associated with modern North Atlantic "long lines," which use several hundred baited hooks attached at varying intervals, was not developed until 1770 by Dutch fishermen. The tenth-century Norman and English fishermen may have removed the cod head, tongue, gills, and gut upon catching; each day's catch was then taken ashore for splitting and crude salting. In the cooler climate of the North Norwegian coast, however, it was cured by sun and air ("stockfish").

Thus we have at this early medieval juncture an essentially seasonal *inshore* fishing regime, even where prosecuted at great distance from a home base in a small coastal community. It was conducted by independent groups of related fishermen who, we surmise, controlled the means of production and largely transported their finished products home for subsistence needs and directly to markets nearby, and around the Lowlands and Baltic coasts.

Middlemen curer-buyers and merchants emerged between the fishermen and their markets in various settings as trade grew around various species.[24] This altered the relationship between fishermen, capital, risk, and profits in often highly uncertain and competitive fisheries. Hence, for example, the rise of Hanseatic merchant associations of North Sea and Baltic coastal towns in the twelfth to fourteenth centuries, and their parallels in various medieval realms. Thus the highly specialized "deep sea" herring fishermen of medieval to eighteenth-century Bohuslän, Sweden, sold or traded their fish in local markets, and to salteries and fish-oil plants operated by merchants from Gothenburg and Uddevolla. The merchants in turn exported these products.[25]

These economic changes signal emergence of an increasingly disadvantaged (economically and politically) maritime peasantry in many areas. Yet the mercantile order in fisheries was fragmented, competitive, and unstable. Uncertain production from fishermen in many scattered locations, and uncertain market returns inhibited undue commitment to production or labor. Likewise, complete dependency upon income from fishing on the part of fishermen was inconceivable in most cases until the Industrial Revolution.

True, fisherman associations or guilds (predecessors, but not parallels to modern unions), such as the thirteenth and fourteenth century Spanish fisherman guilds and federations ("fraternities") of towns and ports,[26] emerged in various coastal areas; as did merchant guilds (e.g., the London Fishmongers of 1154), which increasingly sought to extend the buying function to the fisherman's doorstep because of the difficulty of merchants controlling each other in open markets,[27] a trend which fisherman resistance and restrictive pressures from newly emergent centralized government officialdom (often exploitative in its own right) would impede. But the fisherman as proletarian, a fully committed and dependent labor factor, emerged in large only when navigational and fishing technology, and marketing means, enabled year-round operations to provide for large, daily markets. The rise of the east coast British fresh fish industry in the late eighteenth century is illustrative.[28] Pluralistic, commonly seasonal economic activity, as opposed to fisherman labor specialization, prevailed in most parts of the littoral throughout the medieval and into the contemporary era. The freedom to alternate between fishing and farming, or some other employment activity, e.g., hunting-trapping, logging, trading, migratory construction, and reliance upon producer-controlled, as opposed to non-fisherman buyer- and/or merchant-controlled equipment, has remained a prevalent ideal, if not rational necessity, for many fishermen. It is continuously challenged by interests seeking more control over production and/or profits.

Adoption of a pluralistic economic adjustment among fishermen is only partly explained as a defense against the uncertainty of earnings inherent in what generally remains a *hunting* rather than harvesting enterprise. The steady development of increasingly larger, more mobile and seaworth vessels, and more effective fishing gear and material has only partly subdued this hunting quality. Pluralism frequently expresses a preference for a strong community and home life style, often around a farm. In Norway, for example, where spring and winter herring fishing has long been conceived of as having a "lottery" quality, with the opportunity for remarkable earnings (and failure), the farm and its development was both the aim of fishing participation and economic backup (often, "shock absorber") should one draw a broker in the lottery. It may also be argued that precisely such pluralism helped free many farmers from tenant status and stimulated agricultural growth in nineteenth century western Norway.[29]

Pluralism also expresses an effort to avoid or ameliorate the serious potential drawbacks of labor factor status in capitalistic operations. For example, when relatively capital intensive seine-net and large-boat operations entered the southern district of West Norway after 1820, this challenged the equity characteristic of the earlier drift-net fishing regime. Østensjø writes:

> With the seines, capitalism broke into herring fishing with its advantages and drawbacks, with big catches and large profits for the few owners of boats and tackle and a hard grind and little or no return for the many hired folk or *lottfiskere* (fishermen who depended on a certain share of the catch for payment), the so-called seine-dogs, who had no part in the expensive boats and fishing tackle and worked at miserable wages or uncertain shares. There was more equality in the net parties (*garnlag*), which ordinarily were unions of independent fishermen who had their own nets and a boat in common; but after 1850 the net fishing, too, was partially "capitalised."[30]

Elsewhere, as among contemporary Newfoundland fishermen, pluralism would persist in areas or pockets of otherwise industrialized economies, due to an underdeveloped employment market or economic incentives insufficient to compensate for the economic and social values lost by withdrawal from a pluralistic regime—values too often underestimated by industrially oriented economic policy makers. Clearly, this pluralistic orientation, often erroneously portrayed as "traditional fisherman resistance to change," opposes government mercantilistic and industrial prejudices and policies in the past and present.[31] In other cases, notably Iceland, economic pluralism is an integral component of government policy and economic progress.

Generally, then, European North Atlantic fishing remained and developed as one, albeit major, complex in a seasonally pluralistic economic adjustment which might include, for example, farming, forestry, hunting, trading, and other town and country occupations down to modern times, when migratory construction work became yet another alternative for combination. One could rarely rely upon fish alone, given, at least, its seasonality and uncertain returns. By and large, men undertook the extractive phase, were often assisted by women in their families in the processing phase, and many marketed their own products as independents well into the nineteenth century, by which time mercantile trade had largely assumed the marketing function.

The organization of fishing introduced to the western North Atlantic rim by the end of the fifteenth century was under strong mercantile inspiration and control, and was very much like that in Europe. European fishermen and colonists moved quickly to exploit newly discovered virgin cod and other Northwest Atlantic stocks. They and their technology displaced most coastal Amerind and Eskimo peoples. They came first to Newfoundland waters as seasonal visitors—specialized distant water fishermen operatives lacking the mandate to settle the New World; still others entered as or became fishermen-farmers. Native fishermen hunters and gatherers, and horticultural peoples along the temporate eastern seaboard, were gradually displaced from the coasts. A few remain in small corners of the littoral such as coastal Labrador, New Brunswick, Nova Scotia, and southwestern Greenland, vestiges of their prior occupancy. But they play little part in the aggregate Northwest Atlantic commercial fishery statistic; most of their fishing effort is toward subsistence requirements.

The same Old World adjustments, therefore, were introduced to the eastern Atlantic littoral: farmers, fishermen-farmers, and more specialized, yet largely pluralistic fishermen. Some of the latter settled in, essentially, fishing-station communities (most of which developed small-scale gardening and stock production to supplement their subsistence requirements), as found along the Newfoundland coast. Some, like the eastern or Baltic Swedish fishermen-burghers, settled in growing occupationally diversified port towns like New Bedford and Gloucester, with fishing firms oriented to trade with distant markets "across the sea," then a growing inland agricultural population.

The international competition for access to and control over New World marine and other resources was intense, of course, as the centuries marked by wars over this space between England, France, Portugal, Spain, and the American Colonies testify. But this competition expresses the general developmental processes at issue in this paper: The evolution of new technologies and increasingly complex organizational forms geared to the more effective and intensive exploitation of North Atlantic fishery resources.

INDUSTRIAL FISHERIES AND BEYOND

The myriad paths one may follow in a thorough study of the evolution of North Atlantic fishing are only partially broached in the preceding discussion. The discussion to follow is restricted to two major questions: What is the status of modern North Atlantic fisheries, and what is their future? First, the fisheries of concern here are commercial or industrial.

The rise of modern North Atlantic industrial fishing is roughly assigned to the mid-nineteenth century advent of engine-powered vessels exploiting North Sea whitefish to supply English and mainland European markets. Industrialism may be taken here as increasingly characterized by machine technology, marketing of men's labor, the concentration of workers engaged in single enterprises, the existence of the entrepreneur, a special social type, and rapidly expanding markets.[32] North Atlantic fisheries generally industrialized rather slowly. W. W. Rostow, for example, suggests that the industrialization of British fisheries was about fifty years late; "it reached maturity in about 1900, whereas the British economy as a whole did so in about 1850."[33]

Any one of the characteristics given in the above industrial fishing model provides fertile ground for a lengthy treatise; such is not my objective here. Rather, I am concerned with what is unstated, but implied by the emergence of industrial fishing. It has been predicated upon a natural resource which provided what seemed boundless opportunity for expansion of exploitative activities, with the exception of a few species (e.g., lobster and salmon); common property, requiring no rent, and seemingly inexhaustible until this century.

That various kinds of local and regional regulatory regimes developed around the littoral to control access to fish resources (e.g., in the competition between drift- and seine-net fishing in the 1850s in western Norway)[34] prior to the advent of machine technology, expresses more a recognition of limited local resource exploitative *opportunities*—given existing technology—than the notion that the fish biomass itself was significantly influenced by man's predatory actions.[35] The problem posed was one of balancing local opportunities, space, and fish entering local waters, with numbers of fishing units.

Certainly until the mid-twentieth century, the principal approach or solution to the uncertainties posed by limited local resources was to develop more effective technology capable of going where the fish were. The fish were there to sustain growing effort responsive to growing markets.

Growth of Efficient Technologies

The *general* development of fishing vessels and gear in the historic era evidences continuing efforts to reduce uncertainty in fishing operations and increase surplus production for markets. Larger, more durable and seaworthy vessels, capable of ranging widely, equipped with fish detection and (multipurpose) extraction (and, often, processing) gear enabling operations in a diversity of environments and for a diversity of species, all manifest increasing efficiency in productive technology—given that the resource is there to be found. It is sought after increasingly, rather than awaited, as in trapping operations.

The modern distant water fleets, able to range over many North Atlantic fishing grounds, are somewhat enigmatic in this regard, when compared with middle-range trawlers of coastal nations fishing the same waters. The former may be a "dinosaur" development in the general sense, though

efficient within the narrow interests of a specific country and its industry—
whether socialist or capitalist.

Expanding Use of Natural Resources

Fishing nations have steadily pressed their operational limits into
deeper waters, up to 350 fathoms or more, as their technologies have progressed.
The more common commercial species have been pressed beyond their apparent
sustainable yields; in some cases to the point of commercial "extinction."
Present efforts are increasingly oriented toward development of heretofore
unexploited and underexploited species for which there may even be a market
lacking. Pressed on by growing world protein requirements with a seemingly
insatiable future, fishery specialists now consider all forms of marine life as
potential commercial resources.

Particularly since the Second World War, increasing scientific and
political attention—though hardly enough—has been focused upon the implications of pressing fish resources to their outer sustainable yield limits. The
dramatic expansion in North Atlantic fishing fleets, fishing effort, and production in the last twenty five to thirty years heralded what recent stock declines or
abrupt failures confirm: an imperative need exists for a scientifically sound
and rigorously enforced international fishery managerial regime to assure long-
range stock stability for commercial purposes. And this must be done with
equity, based in parge part upon biological assessments of what fishing stocks
will sustain, and decisions as to who is entitled and able to exploit them. Given
that scientific estimates of maximum sustainable yield levels for various stocks
have predictable inaccuracies of ± 30 percent (a conservative estimate from an
authority involved in making such estimates for the International Commission
for Northwest Atlantic Fisheries) or more, there is reason to expect continuing
pressure toward the upper limits of resource renewal. Hence, little change in
current fish stock "management" practices may be expected on the international
level for the immediate future. We will return to some implications of this
expectation later.

Increasing Occupational Specialization

The long-term shift to ever more versatile and sophisticated fishing
technologies has occurred at considerable capital cost, which, it seems, must be
increasingly justified by year-round fishing operations in many fishing nations,
even where it involves seasonal shifts from one species to another. Such a
technology and fishing regime contributes to development of a specialized
fisherman labor force in most North Atlantic fishing nations.

Particularly since the Second World War, some nations, e.g., Poland,
which lacked a deep sea fishing tradition, had to develop a deep sea fishing
labor force almost from scratch to man new fleets of distant water trawlers.
But recruitment to such fleets is widely fraught with difficulty. The average age

of trawler fishermen increases yearly; and alternative, socially less costly employment lures many potential young fishermen away from the fishing industry in many, if not most, North Atlantic fishing nations.

Against a background of general decline in the fisherman labor force of the North Atlantic since the turn of the century, recruitment to the more capital intensive fleets has been greatly encouraged by higher remuneration, and improved work conditions and benefits. For example, in the Polish fishery, a 1960 report indicated average fishermen earnings might be two and one half times that available in shore occupations.[36] Mechanization has also reduced the labor-intensive character of fishing employment, while simultaneously emphasizing technical skills. Nevertheless, the status of fishermen, perhaps particularly those working for state-subsidized managerial capitalists,[37] is generally low among North Atlantic fishing nations, with the possible exceptions of Iceland, Norway, Sweden, and Denmark. This fact may weigh heavily against present efforts to develop an effective "professional" fishermen labor force in future years.

The part-time or presumably pluralistic fishermen remain an important part of the fishing labor landscape around the littoral, although they seem to be in decline proportionate to the total fisherman labor force. Most regrettably, there is a paucity of good information describing their productive role in detail. This fact cannot be taken lightly as they play an integral role as a labor pool in many fisheries and other maritime enterprises in the rural littoral. Their numbers in, for example, Sweden in 1968 are estimated at 35 percent of the fishermen labor force; about 25 percent in Denmark; 55 percent in Norway in 1966; and 70 percent in Ireland in 1966.[38]

The apparent decline in fishermen pluralists, many of them smallholders as in preindustrial times when they dominated the fishing labor force, is in part a function of the rapid growth in capital-intensive fishing operations. In many cases, the latter have virtually intercepted fish that might have been taken by the seasonal inshore fishermen. Here, we enter upon a dilemma seemingly insoluble in the context of international fishery deliberations, though, in fact, created in part by principles guiding decision-making at this level.

In Newfoundland, for example, the success of more than 80 percent of its 15,000 fishermen depends largely upon fish (especially cod) striking inshore in good quantity in the summer months. The advent of increasingly intensive international and domestic offshore fishing on nearby grounds—and in frequent violation of international limits—in recent years, however, has severely depressed inshore landings. The Labrador and Northeast Coast inshore fisheries, for example, have experienced unprecedented catch failures in recent years to the point that many fishermen have left inshore and nearshore fishing.

Rather than attempt to control offshore extraction to protect the stocks necessary to the inshore fishery, an admittedly formidable undertaking for many reasons not relevant here, provincial and federal governments and buyer-processor-fleet operators have operated to discourage the seasonal fishery in favor of developing a fully committed "professional" fishing labor force. A principal means in their efforts was implementation of a resettlement and centralization scheme to shift scattered coastal populations to "Fishery Growth" centers. Here, they might provide the major source of semi-skilled and skilled labor for industrial trawler fishing and plant operations, and it would be difficult to continue their previous fishing routine.

In consequence, fishermen pluralists are pressured to assume a more fully industrial labor commitment. Expectations that they might become the labor pool to crew expanding trawler fleets were frustrated. Contrary to this rather simplistic economic model of labor fluidity, they have declined to do so. Even the lure of special fishery college training programs has failed to overcome this resistance to date. There is the suggestion that the pressure to industrialize these fishermen constitutes a special kind of "industrial tyranny," in that neither national nor provincial fishery policy-making has been sufficiently subject to influence by fishermen. Certainly the absence of an organized fisherman voice until lately has been contributory, while government policy-makers have fallen in line with views of fleet and plant industrialists committed to capital-intensive operations (like their mainland industrial models) seeking sure, regular supplies for plant processing lines and fresh fish export markets.

Strangely, new management regimes expected from Law of the Sea deliberations will not be a panacea for their problem, as many inshore fishermen may hope. The equation of maximum sustainable yield offers little place for the degree of underfishing necessary to insure relatively less mobile inshore fishermen of reasonable catches. The resource management dialectic here augers for a more fully industrial fishing regime in the future if Canada's potential share is to be increased and taken. What the loss of the fisherman pluralist may mean to rural Newfoundland community and economic life remains an important domestic and provincial question; it is also one to be reckoned with generally around the North Atlantic littoral.

Growth of Complexity in Fishing

This evolutionary overview has taken us rather quickly from subsistence-oriented fishermen of the Paleolithic to modern industrial fishermen occupationals. The former were solely dependent upon individual knowledge, skill, and technical kit applied within the constraints of economic needs and resource availability. The latter is a specialized employee who fares well if he plays but one role well as member of an occupationally diverse team of resource extractors working to supply shore managers and processing lines, which in turn serve often distant foreign markets, whose requirements and fluctuations are shaped by forces largely beyond fisherman control.

The modern fisherman receives a cash return for his labor, though often partly composed of both wages and, as of old, a share of the landings.

Various agencies of local, provincial, and state governments, and international fishery commissions, overlook and, to an increasing extent, shape his fishing efforts. Likewise, various domestic institutional bodies, such as loan boards, banks, and associations—labor, trade, educational—underwrite and determine the development of his fishing technology, his relationship to the means of production, his cognition of resources, and recruitment and careers in fishing.

The diverse fishing adaptations that have emerged, and are still evolving, against this relatively recent background of industrial societal complexity are only fragmentarily understood. For example, the industrial capitalism of the British deep sea or industrial water fishing industry is one variant. That of the nearwater fishery around much of coastal Scotland, where fisherman and familial ownership of vessels has a long tradition, is another. The stress upon capital-intensive fishing, as the Scandinavian fisheries evidence, may be achieved with fisherman-owned and -managed firms and cooperative plant operations, or via the non-fisherman capitalist alternative.

Earlier on I hestitated to represent the North Atlantic littoral as a single European culture area, though one must surely recognize a vast range of similarities among the fishing peoples found here. This is particularly evident on the technological level. Equally impressive, however, is the great range of alternative approaches taken to the organization of fisheries and community and occupational life around them. The delineation and explanation of these alternatives, e.g., in recruitment, ownership, marketing, and fishermen's political roles, and their impiications for human aspirations in this industrial era beckons anthropological study.

Given that "fish are scarce," we are obliged to learn more about how they are taken, by whom, how, and to whom the product is marketed, and the role it plays in meeting human protein needs in a time of global protein insufficiency.

Notes to Chapter 2

1. The peoples of the North Atlantic littoral from the equator to the Arctic are, anthropologically, only somewhat arbitrarily grouped under the general rubric "North Atlantic Maritime peoples." This littoral zone embraces tropical, temperate, and arctic environments, and lacks the singular geographic, historic, or cultural integrity requisite to conceptualization as a (maritime) "culture area" let alone a region (though we may properly view all the peoples of this zone as participants in a Euro-North American industrial culture). With exceptions such as C. Arensberg, "Peoples of the Old World," in *Culture and Community*, edited by C. Arensberg and S. Kimball (New York: Harcourt, Brace and World, 1965), 74-91, who posits an "Atlantic fringe" for European and Old World Culture, anthropologists have approached these peoples rather

narrowly and practically, favoring specific empirical studies—especially of communities. These have been related to lower-level regional, national, or problem-oriented anthropological issues. Speaking especially of fisherman adaptations, anthropologists have only recently begun the difficult task of systematizing their knowledge about the varied maritime cultures of the North Atlantic. See especially R. Andersen and C. Wadel (eds.), *North Atlantic Fishermen* (St. John's: Institute of Social and Economic Research, Memorial University, 1972).

2. A detailed discussion of European fresh and salt water faunal and floral (e.g., shellfish, seals, whales, walruses, seafish, and seaweed) resource uses is provided by J. G. D. Clark in "The Development of Fishing in Prehistoric Europe," *The Antiquaries Journal* 28 (1948): 45-85; and in *Prehistoric Europe: The Economic Basis* (London: Methuen, 1952), pp. 62-90.

3. S. L. Washburn and C. S. Lancaster, "The Evolution of Hunting," in *Man The Hunter*, ed. R. B. Lee and I. DeVore (Chicago: Aldine, 1968), p. 294.

4. G. Clark and S. Piggott, *Prehistoric Europe* (London: Hutchinson, 1965), *passim*; and J. Hawkes, *Prehistory* (New York: New American Library, 1963), pp. 157-58 *et passim*.

5. See Clark, *supra* note 2, p. 48; R. W. Casteel, "Some Archaeological Uses of Fish Remains," *American Antiquity* 37 (1972): 412.

6. B. Salween, "Sea Levels and Archaeology in the Long Island Sound Area," *American Antiquity* 28 (1962): 46-55; B. W. Powell, "Spruce Swamp: A Partially Drowned Coastal Midden in Connecticut," *American Antiquity* 36 (1971): 343-58.

7. Still undetected older evidence of nomadic and perhaps marine fish and mollusk eaters dating back to more than 10,000 B.P. may lie in deeper (up to 60 meters or more) waters far out on the eastern North American continental shelf. This evidence was submerged when sea levels rose following the close of the last glacial period; see K. O. Emery and R. L. Edwards, "Archaeological Potential of the Atlantic Continental Shelf," in *American Antiquity* 31, 5 (1966): 734-37.

8. See the major overview on this subject by E. Rostlund, "Freshwater Fish and Fishing in Native North America," *University of California Publications in Geography* 9 (Berkeley, 1952).

9. G. Gjessing, "Maritime Adaptations in North Norway's Prehistory." Paper presented at the IXth International Congress of Anthropological and Ethnological Sciences meetings, Chicago, 1973, p. 9.

10. Some North Atlantic marine environments were particularly rich and continuously supported large populations from the Mesolithic to the present. For example, single settlements with up to 300 inhabitants are reported on the Arctic Sea coast between Løfoten and Murmansk. See W. Fitzhugh, "The Circumpolar Concept Revisited: Toward Comparative Studies of Northern Maritime Adaptations," p. 17. Paper presented at the IXth International

Congress of Anthropological and Ethnological Sciences meetings, Chicago, 1973.

    11. See Clark and Piggott, *supra* note 4, p. 133, and especially their discussion of "hunter-fishers," pp. 114–34.

    12. Clark suggests, for example, that salmon vertebrae worn as beads found in a burial at Grimaldi, near Mentone above the Italian Mediterranean, were extracted from the Atlantic watershed, and traded by Magdalenians to the Grimaldians. *Prehistoric Europe*, p. 242.

    13. Fitzhugh, *supra* note 10, *passim*.

    14. Clark, *Prehistoric Europe*, 56–58; see also his discussion of "coastal and maritime" catching and gathering, 62–90.

    15. Washburn and Lancaster, *supra* note 3, p. 296.

    16. For the present discussion, this expression is used to embrace both freehold and crofter or tenant fishermen. They are both economic pluralists, though they differ vastly in economic and social position. The development of coastal fishing and reindeer herding particularly in Northern Scandinavia is still another Neolitic ecotype; *Coast Lapp Society II* (Tromsø: Universitetsforlaget, 1965. It is geographically limited to the Old World, and, it seems, exposed to the same basic developmental trends affecting littoral fishing adjustments which became combined with farming or (in the New World) horticulture. My knowledge of this adjustment is limited, however, so discussion is confined to the more common fisherman-farmers.

    17. Clark, *Prehistoric Europe*, pp. 48–51.

    18. J. B. Griffin, "The Northeastern Woodlands Area," in *Prehistoric Man in the New World*, ed. Jesse D. Jennings and Edward Norbeck (Chicago: University of Chicago, 1964), pp. 223–58. On the Powhatans, see B. C. McCary, *Indians in Seventeenth Century Virginia* (Williamsburg, Virginia: Virginia 350th Anniversary Celebration Corporation, 1957).

    19. See, e.g., R. Østensjø, "The Spring Herring Fishing and the Industrial Revolution in Western Norway in the Nineteenth Century," *Scandinavian Economic History Review* 11 (1963): 135–55.

    20. O. Löfgren, "Maritime Hunters in a Peasant Setting: A Comparative Discussion of Swedish Peasant Fishermen," in *North Atlantic Maritime Cultures*, ed. R. Andersen (The Hague: Mouton, in preparation).

    21. J. R. Coull, *The Fisheries of Europe* (London: Bell and Sons, 1972), pp. 60–63 *et passim*.

    22. An elaboration of the history of fishing around specific species is beyond my present purpose, but we may note that the important role of herring fishing arose only recently. Herring remains are remarkably absent from prehistoric European settlements, though the basis of a thriving fishery in some areas by the tenth century. Clark attributes this to the fact that "as with cod nets, manufacture of the drift nets required for herring involved more labor than could be justified by the needs of small communities of farmer-fishers." Nets (presumably *drift* nets) were not used on the Norwegian coast until late in the

seventeenth century, when they are related to supplying large and distant markets. Clark, *Prehistoric Europe*, pp. 89-90.

23. Coull, *supra* note 21, p. 63.

24. M. Gray, "Organisation and Growth in the East Coast Herring Fishing 1800-1885," in *Studies in Scottish Business History*, ed. P. L. Payne (London: Cass, 1967).

25. G. Utterström, "Migratory Labour and the Herring Fisheries of Western Sweden in the 18th Century," *Scandinavian Economic History Review* 7, – (1959), 3-40.

26. See G. M. Orozo, "Fishermen's Guilds in Spain," *International Labour Review* 94, 5 (1966): 465-76.

27. J. Ross, personal communication.

28. See, e.g., J. Tunstall, *The Fishermen* (London: MacGibbon and Kee, 1969).

29. Østensjø, *supra* note 19, pp. 152-53.

30. *Ibid.*, p. 146.

31. See, e.g., Löfgren, *supra* note 20; E. Antler and J. Faris, "Adaptations to Changes in Technology and Government Policy: A Newfoundland Example," in *North Atlantic Maritime Cultures*, ed. R. Andersen, in preparation; O. Brox, *Newfoundland Fishermen in the Age of Industry* (St. John's: Institute of Social and Economic Research, Memorial University of Newfoundland, 1972); C. Wadel, *Marginal Adaptations and Modernization in Newfoundland* (*Ibid.*, 1969).

32. T. Burns (ed.), *Industrial Man* (Harmondsworth, England: Penguin, 1969), p. 7. But cf. Østensjø, *supra* note 19, p. 135, for an alternative formulation.

33. Tunstall, *supra* note 28, p. 19. Industrialization came to eastern North American, and particularly eastern Canadian, fisheries somewhat later. In Newfoundland, for example, dory schooners were still in use on the Grand Banks into the 1940s, by which time the first motor ground fishing trawlers were established in the same ports. Parenthetically, though largely a convenient legal technicality of economic and political service to owners of vertically integrated trawler fleets and fish buyers at large, Newfoundland and Nova Scotian fishermen were classified as "co-adventurers," not "employees," until the early 1970s. A major consequence, trawler fishermen (and their inshore counterparts), who were engaged in year-round fishing for non-fisherman owned plants, were legally prohibited from organizing for collective bargaining. See C. Steinberg, "Collective Bargaining Rights in the Canadian Sea Fisheries: A Case Study of Nova Scotia" (Ph. D. dissertation, Columbia University, 1973). The governments of Newfoundland and Nova Scotia have emended provincial labor laws only in the last few years to enable collective bargaining among fishermen, and labor organization is now in progress.

34. Østensjø, *supra* note 19, p. 146.

35. R. Andersen and G. Stiles, "Resource Management and Spatial Competition in Newfoundland Fishing," in *Seafarer and Community*, ed. P. Fricke (London: Croom Helm, 1973), pp. 44-66. See also "Public and Private Access Management in Newfoundland Fishing," in *North Atlantic Maritime Cultures*, ed. R. Andersen (The Hague: Mouton, in preparation). *World Anthropology* series, Proceedings of the IXth International Congress of Anthropological and Ethnological Science Meetings, Chicago, 1973.

36. Coull, *supra* note 21, p. 124.

37. P. Howden, "The Hull Fishermen and Worker's Control," *Anarchy* 86 (1968), 8 (4), 97-113.

38. Coull, *supra* note 21, p. 125.

Chapter Three

The Regional Consequences of a Global Fisheries Convention

Douglas M. Johnston

FISHERY JURISDICTION ISSUES AT LOS III

Some Anticipated Features of a Future International Fishing Convention

On the basis of fishery proposals submitted by members of the U.N. Seabed Committee up to the fall of 1973, it is possible to describe the current trends of opinion on the fundamental fishery issues to be resolved at the Third U.N. Conference on the Law of the Sea (LOS III). Prediction of the outcome of such a complicated lawmaking conference is a dangerous pastime, but certain features can be anticipated with relative certainty.

There will be signed, or opened for signature, either a single instrument or a number of instruments dealing with fishery jurisdiction issues. Most of these issues are likely to be resolved in the text of a treaty which permits coastal states to establish something in the nature of an economic zone up to 200 miles in breadth. It is possible, but less likely, that the coastal state may be permitted to create such a zone beyond 200 miles out to the farthest edge of the continental shelf (margin), if its geography permits.

One of the major purposes of this kind of instrument would be to grant to the coastal state exclusive fishery jurisdiction within the economic zone. The present preference seems to be for an explicit recognition of the

---

This paper is based in part on a study prepared for the Working Group on Living Marine Resources of The American Society of International Law.

coastal state's "sovereign rights" to the living resources within the zone, and apparently this expression is intended to comprise exclusive managerial authority over the fishery stocks in the area as well as exclusive fishing rights to all species enclosed in the zone.

There also seems to be widespread support for the proposal that the coastal state should be guaranteed some kind of preferential rights to the living resources in more distant areas adjacent to the economic zone, but at the time of writing it is uncertain what is meant by this claim.

Presumably an article will be added urging all states to cooperate with regional fishery commissions as well as global agencies. Such a provision would be interpreted to mean that coastal states should cooperate at least in fishery research with regional fishery commissions which exercise managerial authority over fishing activities in areas adjacent to their economic zone. It might also be interpreted to mean that the commission and the manager coastal state should attempt to maintain similar, or at least compatible, managerial standards and criteria.

It is more difficult to predict how specific a coastal state convention will be regarding the coastal state's obligations toward the existing regional fishery commissions which have jurisdiction at present *within* 200-mile limits. It is possible, for example, that by way of compromise to secure more signatures for the treaty most delegations may be willing to accept a provision which "urges" the coastal manager state to *consult* with such commissions, or in their absence with co-user or neighboring states, with respect to designated aspects of its fishery management policy applicable within the economic zone. In addition, or in the alternative, it is also possible that the coastal manager state may be "urged" to submit designated aspects of its fishery management measures to *review* by the regional commission, or by co-user or neighboring states. It is certainly predictable that most of the distant fishing states at LOS III will press for a provision creating an obligation on the coastal manager state to submit designated kinds of fishery disputes to third party *settlement* in a form to be agreed upon by the parties to the dispute. Such proposals may be acceptable to most coastal states only on condition that the treaty contains provision that, pending the implementation of these procedures of consultation, review, and dispute settlement, the coastal manager state has sole responsibility for the management of the living resources of the economic zone and may, therefore, adopt *interim* measures until differences are reconciled.

Alternative Modes of Treatment

The Sovereign Economic Zone. On the basis of present trends, it seems probable that fishery jurisdiction issues will be resolved in the coastal state's favor by the creation of a multi-functional 200-mile economic zone; or, more properly, by the conclusion of a general convention that recognizes the right of coastal states to create such a zone in accordance with newly prescribed criteria. In most versions of this kind of proposal—including Latin American and Caribbean proposals for the validation of claims to a "patrimonial sea"—the coastal state would be recognized to have "sovereignty" or "sovereign rights" over all living as well as nonliving resources in the zone. Only slightly less emphasis is placed upon the need to recognize the coastal state's exclusive authority in the zone over the control of marine pollution and the regulation of scientific research.[1]

In its unmodified form the economic zone approach would impose minimal restrictions on the coastal state's exercise of its fishery management authority and no restraints at all on its right to exclude foreign fishing within the limits of the zone. The advantages of simplicity and uniformity are regarded by most economic zone protagonists as outweighing the cost of ignoring the different behavioral characteristics of species which is inherent in an unmodified zonal approach to fishery jurisdiction issues.

The Nonsovereign Economic Zone Approach. This approach, aimed at the modification of existing proposals for an unqualified 200-mile economic zone, would offer a compromise not only with the more moderate distant fishing interests but also with the functionalist logic of the species-by-species approach to fishery management supported by some coastal states, such as Canada and the United States. This approach might be characterized as "eclectic," designed to secure the advantages of the spatial (zonal) approach without losing all the virtues of the functionalist approach to fishery management. According to the latter, there should be different modes of treatment for sedentary, anadromous, (other) highly migratory, and "coastal" species, varying with the life-cycle and behavior of these groups.

It is believed, however, that the species-by-species approach, as described in recent Canadian[2] and U.S. working papers,[3] may be too complicated to be salable to developing countries with limited confidence in their own scientific capability. Moreover, it may prove impossible to persuade a two-thirds majority at LOS III to accept the view that the state of origin, as manager state, should be alone entitled to the capture of *anadromous* species, like the Atlantic salmon, which spend most of their life-cycle in the high seas, feeding and growing in waters superjacent to the shelf of other states. Apart from other objections, a duty on other states to abstain from harvesting such species would be regarded as a denial of their "sovereign rights" to the marine resources within the proposed economic zone.

On the other hand, it appears there may be a growing recognition of the need to deal with *migratory* species on a separate basis. In a recent working paper cosponsored by Canada, Kenya, Senegal, India, Sri Lanka, and Malagasy,[4] a distinction is made between species which have *limited migratory* habits; but these species are singled out for purposes of management *beyond* the coastal state's exclusive fishing zone (which is deemed to be coextensive with its economic zone) in areas where it is suggested the coastal state should have *preferential* rights of capture. This proposal, then, though couched in the language of the species-by-species approach, is apparently motivated by the desire to extend beyond the economic zone the concept of the coastal state's special fishing rights, rather than by the desire to place managerial controls on a scientifically rational basis. The merit of this apparent concession to the functionalist logic of the species-by-species approach is placed in doubt by three other features of the working paper: its failure to provide for international managerial controls over highly migratory species *within* the coastal exclusive fishing zone (or economic zone); its failure to provide for a separate regime in respect of anadromous species, treated *sui generis*; and its failure to specify what the spatial limits of exclusive fishing rights should be.

A Separate Fishing Convention.   A separate LOS III treaty, dealing solely with fishery matters, would be the most consistent with the functionalist approach to law of the sea issues. To the scientist, it represents the most logical method of treating fishery resources, to the extent that this is severable from the rational management of the coastal marine environment in general. In addition, a separate fishing convention is legally ideal, since it could deal more comprehensively with fishery problems and provide a juridical link between the fishing rights and conservation responsibilities of states, both coastal and non-coastal. At present, however, there is relatively little interest among states preparing for LOS III in a treaty which would set out the suggested principles of fishery management. One suspects that many delegations, especially from the developing regions of the world, would feel at a disadvantage in the process of preparing a draft of this kind.

A comprehensive fishing treaty might include, *inter alia*, the following principles.

a.   Subject to (b) below, the coastal state is the sole manager of the living resources beyond its 12-mile territorial limits out to a line 200 miles seaward of the baseline of its territorial sea or to the edge of the continental shelf (margin), whichever is the greater.

b.   With respect to anadromous species, the managerial authority of the "feeding" coastal state is displaced by that of the state-of-origin. With respect to other highly migratory species, the managerial authority of the coastal state outside its 12-mile territorial limits yields to the primary authority of an international (regional or oceanic) management commission whose jurisdiction is limited to particular stocks or species under the "highly migratory" category.

c.   With respect to all species over which it has sole managerial authority, the coastal state has exclusive fishing rights but may, of course, permit foreign fishing, if it chooses, subject to managerial controls and conditions of entry prescribed by it.

      d. Where the nature or scope of the coastal state's managerial authority is contested by an international fishery commission, the former is *urged* to enter into negotiations with the latter in order to reach a regional accommodation consonant with the spirit of the (LOS III) fishing treaty and with the general principles of rational fishery management. The commission in this situation is *required* to comply with the coastal state's request to enter into negotiations to that end.

      e. Where foreign fishing exists on a commercially significant scale within the new jurisdictional limits of the manager coastal state, as recognized in this treaty, the limitation of foreign fishing effort should be sought only on a "phase-out" basis in order to avoid serious and abrupt economic dislocation.

      f. Fishery management policy and practices, whether adopted by coastal state, by state-of-origin, or by international commission, are subject to dispute avoidance and dispute settlement procedures in order that management decisions can be made and implemented in an environment free of unnecessary conflict among the states affected.

      g. In the making of fishery management decisions, the coastal state (or the state-of-origin) is *urged* to consult with neighboring states whose fishery interests might be affected adversely by such decisions, or with any international fishery commission with managerial authority over areas immediately adjacent to the jurisdictional limits of the coastal state (or state-of-origin).

      h. In the making of fishery management decisions, any international fishery commission is *urged* to consult with nonmember states whose fishery interests might be affected adversely by such decisions, and is *required* to give full weight to objections raised by adjacent coastal states which are consonant with the spirit of this treaty and with the general principles of rational fishery management.

      i. In the implementation of fishery management decisions, the coastal state (or the state-of-origin) is *urged* to comply with mutually acceptable review procedures at the request of neighboring states whose fishery interests might be affected adversely by such decisions, or at the request of any international fishery commission with managerial authority over areas immediately adjacent to the jurisdictional limits of the coastal state (or state-of-origin).

      j. In the implementation of fishery management decisions, any international fishery commission is *required* to comply with mutually acceptable review procedures at the request of adjacent coastal states whose fishery interests might be affected adversely by such decisions.

      k. In the event of a dispute over fishery management practices between a coastal state (or state-of-origin) and a neighboring state or international fishery commission, any one party to the dispute is *required* to comply with a request by the other to submit the issue to settlement in a mutually acceptable manner, failing which in a manner to be decided by a designated impartial person, e.g., the Director-General of Fisheries, FAO.

A Multiple-Treaty Regime.   The fourth alternative approach to fishery jurisdiction issues would be by the general acceptance of a new coastal regime based on a general convention and a series of subordinate and interdependent "functional" treaties.

The "parent" convention would establish a multipurpose *coastal maritime regime*, under which the coastal state would acquire *special* rights— some exclusive, others preferential but nonexclusive—to the resources and environment of coastal maritime areas, as defined in the "derivative" conventions. This *constitutive* instrument would affirm the principle of the coastal state's "special interest" in the resources and environment of coastal maritime areas, thereby defining as specifically as possible the limitations on the complementary principle of the freedom of the high seas. This instrument would refrain from specifying the *spatial extent* of the applicability of the principle of the coastal state's "special interest." In other words, it would be limited to establishing the *legitimacy* of the coastal maritime regime, leaving the question of its *scope* to be dealt with in the "derivative" instruments, described below. It would, however, attempt to define the *nature* of "special interest" and identify the institutions, procedures and criteria to be used in the avoidance or settlement of disputes regarding the nature of "special interest" under the "parent" convention and the scope of "special interest" zones established unilaterally or in accordance with the "derivative" conventions.

Each of the "derivative" conventions would grant to the coastal state the right to establish a single-purpose zone for the exercise of exclusive or special rights and duties. There would be one for *mineral resources*, another for *living resources*, and a third perhaps for *environmental control* (marine pollution prevention and the regulation of marine scientific research). It is likely that at least in the first of such instruments, the parties would be able to agree on uniform maximum limits (say, 200 miles). In the second, perhaps the coastal state would be given sole managerial authority out to the 200-mile limit or to the edge of the continental shelf (margin), whichever is greater; but more difficulty would perhaps be encountered in defining maximum limits for exclusive and preferential fishing rights. The determination of spatial limits of coastal state jurisdiction might be even more difficult in the third "derivative" convention, but many states would no doubt settle for 200 miles.

CLASSIFICATION OF FISHING STATES

It is useful to classify all fishing states in two ways: first, by reference to their current fishery interests, especially as reflected in their general attitude to current jurisdictional issues in the law of the sea; and second, by reference to the treaty rights and obligations they have acquired through signing, ratifying, or otherwise adhering to fishery agreements and fishing conventions.

Fishery Interests

The first method of classification is one that requires a multi-factoral approach to the identification of "national interest" in the use of sea fisheries. It might seem that national interest would be accurately reflected in positions adopted and advocated by government delegations in the course of preparations for LOS III, but often these diplomatic statements are designed to serve tactical purposes and give a distorted impression of government policy. Sometimes a line of argument or claim is introduced in order to probe an area of possible compromise or to signal where trade-off concessions might be negotiated. More frequently, an exaggerated position is maintained for reasons of bargaining psychology. Diplomatic statements made in the course of conference preparations are a sufficiently unreliable evidence of government policy that they must be supplemented by other kinds of evidence of national interest: geographical, economic, political, and legislative. With a general approach of this kind one can construct with reasonable accuracy actual or presumed national "positions" on basic jurisdictional issues in the law of the sea. Specifically, one can categorize states in this way with respect to their "positions" on current proposals for the extension of coastal state jurisdiction over fishing and fishery conservation.[5]

Since the most familiar current proposals identify the 200-mile limit of the *economic zone* (EZ) as the maximum limit of coastal state fishery jurisdiction, one can classify states in accordance with their reaction to this kind of claim, which is such a radical innovation in international law as to divide the world community into strongly opposed groupings. Accordingly, those states which are in favor of EZ-type proposals may be described as *protagonist* states. Those states opposed may be referred to as *antagonist* states. States which have yet to indicate their preference one way or another may be designated as *uncommitted*. Last, states which still have some difficulty in determining their final position on such proposals may be categorized as *equivocal*. In Table 3-1, the twenty-five states with significant fishery interests in the North Atlantic north of the tropic of Cancer (excluding the Baltic, Mediterranean, and Black Seas) have been classified in the above way. In column 1 each of these states is given a rating of intensity on the scale from PRO 3 (very strong proponent of a 200-mile economic zone) to ANT 3 (very strong opponent of such a proposal). By my reckoning, nine of the twenty-five could be characterized as ANT 3, whereas only four could be regarded as belonging to the opposite extreme of PRO 3. Altogether nineteen of the twenty-five are regarded as unfavorably disposed to the EZ concept.

More properly, the reflection of LOS III fishery interest should be looked for in the general attitude of each state towards specific fishery proposals.

Table 3-1. North Atlantic Fishing States: Diplomatic Attitudes and Treaty Commitments

	1974 LOS III Attitudes		1958 LOS I Commitments			Some North Atlantic Fishery Commitments						Law of Treaties
	200-mile E.Z.	200-mile E.F.Z.	High Seas	Fishing + Conservation	Continental Shelf	ICNAF	NEAFC	Atlantic Tuna	1969 London Convention	1967 Convention on Conduit	1964 ICES	Vienna Convention
Regional States												
Bahamas	PRO 2											
Belgium	ANT 1	ANT 2	A	A	A		P (Res)		P		P	
Bulgaria	ANT 3	ANT 3	A	S	S-R	P					P	A
Canada	PRO 3[1]	PRO 3[1]	S	S	S	P		P	P		P	S
Cuba	PRO 3	PRO 2	S	S	S			P			P	
Denmark	ANT 1	PRO 1	S-R	S-R (Res)	S-R	P	P	P	P		P	S
France	ANT 1	ANT 1	S	S-R	A (Res)	P (Res)	P	P	P		P	
Germany, E. (GDR)	ANT 3	ANT 3										
Germany, W. (FRG)	PRO 3	PRO 3	S	S	S	P	P			P	P	S
Iceland	ANT 1	PRO 1	S	S	S	P	P		P		P	S
Ireland	ANT 1	PRO 1	A			P	P				P	
Italy	ANT 2	ANT 2	A		S-R			P			P	S-R
Morocco	PRO 3[2]	PRO 3[2]	S-R	S-R	S-R		P (Res)	P				
Netherlands	ANT 3	ANT 3	S-R	S-R	S-R	P	P			P	P	
Norway	ANT 2	PRO 1[3]	S-R		A	P	P	P	P		P	A
Poland	ANT 2	PRO 1[4]	S-R	S-R	S-R	P	P				P	
Portugal	ANT 1	PRO 1	A	A (Res)	A	P	P	P		P (Res)	P	S
Romania	PRO 1	PRO 1	A		A	P (Res)	P					
Spain	ANT 3	ANT 3[5]	S-R	S-R (Res)	S-R	P	P	P	P		P	A
Sweden	ANT 2	ANT 2	S-R	S-R (Res)	S-R	P	P	P	P		P	S
U.K.	ANT 3	PRO 2	S-R		S-R	P	P	P	P	P	P	S-R
U.S.A.	ANT 3	ANT 2	S-R		S-R	P	P	P	P	P	P	S
U.S.S.R.	ANT 3	ANT 3	S-R		S-R							
Extra-regional States												
Japan	ANT 3	ANT 3	S-R			P		P			P	S
Korea, S.	ANT 2	ANT 2						P			P	

NOTES: The reference here is to that part of the Atlantic Ocean north of the Tropic of Cancer, excluding the Baltic, Mediterranean, and Black Seas.

1. The Canadian position supports claims to a zone of functionally defined national jurisdiction out to the edge of the continental shelf (margin) or to a line 200 miles seaward from the baseline of the territorial sea, whichever is the greater.

2. By a decree of March 1973, Morocco claims an exclusive fishing zone up to 70 miles in width.

3. Norway's shift in favor of extensive coastal fishing zones is subject to the requirement that accommodations would have to be made with existing regional commissions.

4. This attitude is inferred from Poland's willingness to engage in joint ventures with coastal states claiming extensive national jurisdiction over offshore stocks.

5. Spain has protested vigorously against Morocco's extensive fisheries zones promulgated in 1973 which come within 6 miles of the Spanish coast between Cadiz and Malaga, but this issue has recently been resolved in a bilateral agreement between the two states.

Since exclusive fishing zone proposals have been outnumbered and overshadowed by EZ-type proposals in preparations for LOS III, it is difficult in most cases to do more than speculate or infer what national positions might be on a proposal for, say, a 200-mile exclusive fishing zone (EFZ). In the second column of Table 3-1 the same twenty-five states—the major North Atlantic fishery interests—are classified in accordance with this second, hypothetical test. It is suggested that if the non-fishery components of the 200-mile EZ were excluded from consideration the pattern would vary appreciably; namely, that only thirteen would be opposed to the concept of a 200-mile EFZ. In some cases the national position is expected to be very different: especially, that of the United States, Norway, and Poland, and possibly Denmark and Ireland.

Treaty Commitments

The second method of classification is by reference to national acceptance or nonacceptance of treaty rights and obligations.

Some idea of the pattern of fishery-related treaty commitments in the North Atlantic is provided in Table 3-1. This is not, of course, an exhaustive listing: it merely shows the pattern for these twenty-five states with respect to nine important fishing or fishery-related conventions. The first group—columns 3 to 5—shows the pattern of signature (s), ratification (R), or accession (A), with or without reservation (Res). The second group—columns 6 to 11—shows which of the listed states are parties (P) to the six designated regional conventions. The last column shows the pattern of signature, ratification, and accession with reference to the 1969 Vienna Convention on the Law of Treaties.

The significance of these patterns of treaty commitments in the North Atlantic is discussed in subsequent sections of this chapter.

CLASSIFICATION OF FISHING REGIONS OF THE WORLD

Present trends seem to indicate that a two-thirds majority at LOS III will support most of the anticipated features of an international fishing convention, outlined above. But a bare two-thirds majority vote of approval would not be enough to provide a new global basis for the international law of fisheries. If these elements were put into something like the current EZ package, they would be resisted by most of the states which are characterized at present as antagonist toward such proposals. Since these states include many of the major fishing states in the world, an outcome of this kind would be no solution at all. If LOS III culminated instead in an "eclectic" or modified EZ agreement, the position would be slightly better, but even in this case there would emerge three distinguishable kinds of fishing regions in the world, varying in accordance with the mix of national attitudes in the region toward a LOS III agreement on 200-mile limits of exclusive coastal fishery jurisdiction. The three types of fishing regions would be:

       a. regions which tend to be dominated by the fishery interests of protagonist states ("coastal fishing regions"), e.g., Southeast Pacific Ocean, western Indian Ocean;

       b. regions which tend to be dominated by the fishery interests of antagonist states ("inclusive fishing regions"), e.g., Northeast Atlantic Ocean, Mediterranean and Black Seas; and

       c. regions which are characterized by an acute conflict between coastal and noncoastal fishery interests ("fishing conflict regions"), e.g., Northwest Pacific Ocean, Southeast Atlantic Ocean, Northwest Atlantic Ocean.

Admittedly some of these designated regions, which are based arbitrarily on the breakdown of the oceans for the collection of FAO fishery statistics, might be reduced themselves to subregions, since marine resource tensions exist at these levels in significant degrees of intensity.[6] But in many areas of the world today patterns of alignment on fishery jurisdiction issues are emerging at regional (or macro-regional) levels. The above threefold classification of fishing regions seems useful at least for the purpose of appraising the prospect of serious treaty law conflicts after LOS III.[7]

It should be noted that by this method of classification the North Atlantic will consist after LOS III of two distinguishable types of fishing regions: the Northeast Atlantic, characterized as an "inclusive fishing region," and the Northwest Atlantic, characterized as a "fishing conflict region." The typology suggests that the probable impact of LOS III in the two North Atlantic regions should be analyzed separately.

PROBABLE IMPACTS OF LOS III IN NORTHEAST ATLANTIC FISHING REGION

The Northeast Atlantic fishing region is regulated by a network of multilateral and bilateral treaties. The most important recent fishery treaty development in the region was the establishment of the North-East Atlantic Fisheries Commission (NEAFC) in the North-East Atlantic Fisheries Convention, signed at London in January 1959.[8] This convention was signed by fourteen states belonging to the region: Belgium, Denmark, France, the Federal Republic of Germany (F.R.G.), Iceland, Ireland, the Netherlands, Norway, Poland, Portugal, Spain, Sweden, the U.S.S.R., and the United Kingdom. Most of the parties to this convention, and to the other fishery agreements in the region, are opposed to the concept of a 200-mile economic zone: according to Table 3-1, twelve of the above fourteen are antagonistic toward the EZ. According to the same estimate, only a bare majority, eight of the fourteen, may be opposed to the concept of a 200-mile EFZ by the time LOS III gets under way at Caracas in June 1974. But this is misleading, for of the six suggested EFZ supporters only Iceland can be regarded as an ardent protagonist. The other five—Denmark, Ireland, Norway,

Poland, and Spain—are unlikely to accept a LOS III convention which seeks to grant Northeast Atlantic coastal states the right to declare a 200-mile EFZ without regard for existing regional treaty arrangements. In sum, the region is likely to be "inclusive" in the sense that most of the states fishing there will continue to operate the existing *regional treaty system* which supports the traditional policy of inclusive fishing beyond modest territorial limits subject to commonly accepted conservation restraints on the freedom of fishing.

As suggested above, there are two other alternative outcomes of LOS III with respect to fishery jurisdiction issues: a separate fishing convention and a multiple-treaty regime. But most of the Northeast Atlantic states have been remarkably unimaginative in their approach to LOS III, perhaps inhibited by the emotions flowing for and against sovereign and nonsovereign EZ concepts. At any rate there is little current evidence that they are able and willing to organize themselves as a bloc to pursue either of these routes to realistic objectives of compromise diplomacy.

At the time of writing, the disputes over Iceland's right to an exclusive 50-mile fishing zone have been cooled down by political accommodations, but the British government apparently intends to continue proceedings against Iceland in the International Court of Justice.[9] In the aftermath of LOS III, Iceland will presumably be able to point to a treaty development of some kind in support of the argument that its claim is endorsed in principle by a substantial number of states. In that situation it will be able to assert that its own position on the permissible limits of national fishery jurisdiction is entirely consistent with that of the world majority, though not perhaps with that of the majority of states within the Northeast Atlantic region. This scenario would raise the fundamental question whether a regional consensus, manifested in a regional treaty network and acted out in state practice over many years, could prevail within the region over emerging norms in customary international law supported by a large world majority. Conceivably this battle could be fought on the ground sketched out in Article 64 of the Vienna Convention on the Law of Treaties which provides that:

> If a new peremptory norm of general international law emerges, any existing treaty which is in conflict with that norm becomes void and terminates.

The alleged norm in question would be the newly emerged right of coastal states to establish exclusive and/or preferential fishing rights in accordance with criteria recognized in a law-making convention adopted by a large majority of states. This will be a difficult argument for Iceland to maintain in light of Article 53 of the Vienna Convention, which defines a peremptory norm of general international law as "a norm accepted and recognized by the international community of states *as a whole*." In these circumstances, after the signing of a coastal state convention of some kind, Iceland might seriously consider the advantages involved in adopting a unilateral policy and denouncing the NEAFC Convention in accordance with Article 17 of that instrument.

It should be noted, however, that Iceland did not sign, and has not acceded to, the 1964 London Fisheries Convention, and therefore the question does not arise for Iceland whether, under Article 62 of the Vienna Convention, a fundamental change of circumstances may be invoked as a ground for withdrawing from a treaty which establishes a fishery zone boundary.[10] This difficulty may be very real, however, for certain parties to the 1964 Fisheries Convention, such as France or Spain, that might be induced to accept a compromise version of an international fishing convention. Under Article 15, the 1964 Fisheries Convention is of unlimited duration, and cannot be denounced within twenty years of its entry into force and even then the denunciation cannot take legal effect until two years after notice in writing.

It should also be noted that several of the Northeast Atlantic fishing states have either signed or ratified the 1958 Convention on Fishing and Conservation of the Living Resources of the High Seas (the "Geneva Fishing Convention"), and of these at least six may decide not to sign a future international fishing convention with the features anticipated in this paper: viz., Denmark, Finland, Ireland, the Netherlands, Portugal, and the United Kingdom. Moreover, it is conceivable that a few of the Northeast Atlantic nonsignatories of the Geneva Fishing Convention, such as the Soviet Union, Poland, East Germany, and Romania, might decide to accede to that convention as a kind of rear guard action in the face of growing support for the future international fishing convention likely to emerge from LOS III. In this eventuality it is possible, if not probable, that a majority of the Northeast Atlantic states could point to the Geneva Fishing Convention as the global umbrella under which the regional fishery treaty system of the Northeast Atlantic continues to operate, regardless of the applicability of a future fishing convention to other regions. If most other regions in the world come under the global umbrella of the future international fishing convention, the novel objection might be raised that the Geneva Fishing Convention is no longer universal in character but only regional at best. This would be just one aspect of the difficult problem concerning the effect of a future international fishing convention which purports to repeat or amend the fishery provisions in the Geneva Fishing Convention, the High Seas Convention, and Continental Shelf Convention, if many states decline to accept the former and still adhere to the latter. How many "surviving" ratifications of the 1958 Geneva Conventions are needed to keep them in force? Do their provisions still bind signatories which have declined to sign the future convention? Can a global lawmaking convention change its character in such circumstances and become a binding regional agreement?

It should be noted that the Geneva Fishing Convention, for example, does not contain any provisions for withdrawal, but Article 20 provides:

1. After the expiration of a period of five years from the date on which the Convention shall enter into force a request for the revision of this Convention may be made at any time by any contracting party by means of a notification in writing addressed to the Secretary-General of the United Nations.

2. The General Assembly of the United Nations shall decide upon the steps, if any, to be taken in respect of such a request.

Presumably this provision would enable a counter-strategic move by parties to the Geneva Fishing Convention who have since become parties to a LOS III convention to have the General Assembly initiate procedures for revision of the Geneva Fishing Convention so as to make it compatible with the later instrument. No doubt the actions of the General Assembly would be influenced by the fact that two-thirds of its membership is composed of protagonist states who would presumably have adopted the new convention.

PROBABLE IMPACTS OF LOS III IN NORTHWEST ATLANTIC FISHING REGION

The chief institutional feature of this region is the existence of an international fishery commission, which is now developing a complicated system for the allocation of benefits.[11] The commission is, of course, the International Commission for the Northwest Atlantic Fisheries (ICNAF) established by treaty in 1949. By the allocation system, the convention waters are divided into panel areas; catch limits are prescribed annually for the designated stocks of commercial importance in each area; and it is determined annually how the total quota for each stock in the following year shall be allocated among members of the appropriate panel and other (third-party) states. These annual sharing decisions have become increasingly favorable to the coastal state's claim of privilege or "special interest."[12] In light of the trend toward a considerable extension of coastal state jurisdiction, the key question is how much further the noncoastal ICNAF states will be prepared to go, after LOS III, in accepting coastal state demands for the phasing-out of their fishing operations in Convention waters within the limits of national fishery jurisdiction set out by treaty at the conference.

Since these recent national quota developments are the result of threats by Canada and the United States that they might withdraw from ICNAF, these two coastal states cannot complain that their regional diplomacy has been entirely unsuccessful. Certainly the adoption of any LOS III convention of the types envisaged in this paper would enable the two North American states to bring added moral and legal force to their contention that, as coastal states, they should have exclusive or preferential fishing rights, and sole or special managerial authority over the stocks, in extensive areas of the ICNAF Convention waters. Renewed threats of withdrawal by Canada and the United States after LOS III would be the more credible because they could be executed under the ICNAF Convention with only six months notice to the other parties.

Since the convention waters are divided into separate panel areas, it might prove possible for Canada and the United States to persuade the non-coastal ICNAF states to phase out their fishing activities in those panel areas relatively close to the North American coastline, and to press only for additional preferential rights in the more distant panel areas. Fishery diplomacy of this kind would be given added urgency if the coastal states attached a reservation to their ratification of a LOS III convention in terms which would exempt the application of the convention to the Northwest Atlantic region for a reasonable period of time, long enough to accomplish diplomatic compromises within the regional organization. Even better perhaps would be a provision in the LOS III convention which permitted parties to the convention to enter into a separate agreement to modify the convention as between themselves, in accordance with Article 41 of the Vienna Convention on the Law of Treaties.[13]

Since the present Canadian position on fishery jurisdiction is in line with the prevailing trend toward a sovereign or nonsovereign economic zone, the feasibility of either of the above suggestions—exemption or modification—seems to depend mostly on the flexibility of the U.S. approach to jurisdictional issues. If the United States were willing to make fishery-related concessions to the "protagonist" states in a proposal for a separate fishing convention or a multi-treaty regime, the "protagonist" and "antagonist" states alike might be willing to accept an exemption or modification clause. Either way, it is assumed that Northwest Atlantic regional fishery diplomacy can be successfully conjoined with global law of the sea diplomacy at LOS III, so that the final coastal state convention would not give rise to unforeseen problems in Northwest Atlantic regional fishery diplomacy after the conference. No doubt, the future of fishery diplomacy in the Northwest Atlantic during and after LOS III will be shaped very largely by the acceptability of *joint venture arrangements* by the agencies and enterprises of noncoastal states which are willing to accept extensive coastal state jurisdiction as a kind of security for their investment in long-range fishing capability.

With respect to fishery management, on the other hand, it may be difficult to secure Northwest Atlantic regional compromises if the coastal states are not willing to accept procedures of consultation, review, and dispute settlement. Amendments to the ICNAF Convention would be needed to prescribe specifically in what circumstances Canada and the United States, as the manager states, would be required to involve the ICNAF apparatus in certain aspects of regional fishery management. It might be especially important to have new machinery established for the settlement of disputes regarding the nature and extent of the coastal state's managerial authority.

Difficult treaty law issues are likely to arise out of the enactment of new amendments to the ICNAF Convention. For example, one or two ICNAF member states might decline to be phased out from one of the suggested panel areas in which the coastal state is seeking to acquire exclusive rights and might find it expedient to denounce a part, but not the whole, of the ICNAF Convention, raising questions about the separability of the provisions of that treaty and the applicability of Article 44 of the Vienna Convention on the separability of treaty provisions. Alternatively, a small member state of ICNAF which refused to be phased out of a particular panel area at the insistence of a coastal state might find itself subjected to confiscation practices which would eventually force it to comply with the coastal state's exclusivist policy, so that, were the compliance translated to treaty form, there might be a question later whether that instrument is void under the rule of Article 52 of the Vienna Convention on the legal effects of coercion of a state by the threat or use of force. In this case it would have to be proved that the threat or use of force was "in violation of the principles of international law embodied in the Charter of the United Nations." This in turn would raise questions about the legality of a provision in a U.N.-sponsored international fishing convention which would permit coastal states to resort to force in order to secure their interest in the economic or fishing zone adjacent to their territorial sea.

Notes to Chapter 3

1. For an analysis of EZ-type proposals, see Douglas M. Johnston and Edgar Gold, *The Economic Zone in the Law of the Sea: Survey, Analysis and Appraisal of Current Trends*, Occasional Paper No. 17 (Law of the Sea Institute, University of Rhode Island June 1973).

2. Doc. A/AC. 138/SC.II/L. 8 (July 27, 1972).

3. Doc. A/AC. 138/SC.II/L. 20 (April 2, 1973).

4. Doc. A/AC. 138/SC.II/L. 38 (July 16, 1973).

5. It has been suggested more specifically that for any particular stock of fish four different sets of national interest might be distinguished. See Francis T. Christy, Jr., *Alternative Arrangements for Marine Fisheries: An Overview*, Paper No. 1, Program of International Studies of Fishery Arrangements, Resources for the Future, Inc., (May 1973). On the methodology of measuring "national marine interest," see Lewis M. Alexander, "Indices of National Interest in the Oceans," *Ocean Devel. and Int'l. L.J.* 1, 21 (1973).

6. See, for example, Douglas M. Johnston, "Development, Environment and Marine Resources in the North Pacific," in *Proceedings of International*

Conference ("Asia and the Western Pacific: Internal Changes and External Influences") held at Canberra, April 14-17, 1973. [Publication pending.]

7. See Douglas M. Johnston, "Some Treaty Law Aspects of a Future International Fishing Convention," prepared for Working Group on Living Marine Resources, Panel on the Law of the Sea, The American Society of International Law. [Publication pending.]

8. For a recent appraisal, see Jon Thormodsson, "Some Legal Aspects of the Conservation of Fish Stocks in the North-East Atlantic Ocean," 26 *Úlfljótur* No. 2, Supplement (1973). For a comparison of the functions and powers of NEAFC with those of other international fishery organizations, see Albert W. Koers, *International Regulation of Marine Fisheries* (London, 1973), pp. 171-228.

9. For a thorough analysis of the dispute prior to the political agreement, see Richard B. Bilder, "The Anglo-Icelandic Fisheries Dispute," *Wisc. L. Rev.* 37 (1973).

10. Under this Convention [U.N.T.S. Vol. 581, No. 8432], each signatory state is recognized as having "the exclusive right to fish and exclusive jurisdiction in matter of fisheries within the belt of six miles measured from the baseline of its territorial sea" (Article 2). Moreover, beyond that zone out to the 12-mile limit "the right to fish shall be exercised only by the coastal State and by such other Contracting Parties, the fishing vessels of which have habitually fished in that belt between 1st January, 1953, and 31st December, 1962" (Article 3).

11. For a view that the system is too complicated and inflexible, see Francis T. Christy, Jr., "Northwest Atlantic Fisheries Arrangements: A Test of the Species Approach," *Ocean Devel. and Int'l. L.J.* 1, 65 (1973).

12. At present these sharing arrangements conform approximately with the following guidelines: 40% of the total allowable catch is based on percentages taken by the participating states in the past *ten* years; another 40% is based on percentages taken in the past *three* years; 10% is granted on a preferential share to the nearest coastal state; and the remaining 10% is reserved for newcomer ("third party") states and special conditions.

13. Article 41 of the Vienna Convention on the Law of Treaties provides that:

> 1. Two or more of the parties to a multilateral treaty may conclude an agreement to modify the treaty as between themselves alone if:
> (a) the possibility of such a modification is provided for by the treaty; or
> (b) the modification in question is not prohibited by the treaty and:
>   (i)   does not affect the enjoyment by the other parties of their rights under the treaty or the performance of obligations;

>   (ii)   does not relate to a provision, deregation from which is incompatible with the effective execution of the object and purpose of the treaty as a whole.

2. Unless in a case falling under paragraph 1(a) the treaty otherwise provides, the parties in question shall notify the other parties of their intention to conclude the agreement and of the modification to the treaty for which it provides.

Chapter Four

Implications of Alternative Regimes for the Northeast Atlantic Treaty System

Albert W. Koers

INTRODUCTION

The Northeast Atlantic Ocean is not only one of the oldest fishing areas of the world, but one of the world's most productive as well; its 1970 catch of 10.5 million tons was surpassed only by the Northwest Pacific Ocean and the Southeast Pacific Ocean.[1] The Northeast Atlantic Ocean accounts for about 14 percent of the total world catch, which compares with about 6 percent for the Northwest Atlantic Ocean.[2] However, it would be erroneous to conclude from these impressive figures that all is well with the fisheries of the Northeast Atlantic Ocean. In fact they suffer from the typical problems of most mature high seas fisheries: overfishing and overcapitalization. Although there are still some underexploited species in the area,[3] all highly valuable stocks are heavily fished, while many stocks are being overfished; North Sea herring, Atlanto-Scandian herring and northern cod are merely examples. Overcapitalization, too, occurs in virtually all major Northeast Atlantic fisheries. As a result, too much fishing effort is applied in relation to the catch—the same catch could be produced with less effort and, thus, at a lower cost.

Overfishing and overcapitalization indicate that the Northeast Atlantic treaty system did not achieve optimum results. The treaty system also failed to prevent resort to unilateral action, in some cases by coastal states; Iceland's extension of its exclusive fisheries waters up to 50 miles from its coast provides a recent example.[4] Apparently, the treaty system has been unable to accommodate the interests of these states in a satisfactory manner. These various problems raise the question of which alternative regimes could replace the present Northeast Atlantic fisheries arrangements and what the implications

53

would be of such alternative regimes. This question must also be asked in view of the current developments in the law of the sea generally. It is, for example, very unlikely that the Northeast Atlantic fisheries treaty system would not feel the impact if the 1974 United Nations Law of the Sea Conference were to decide that all coastal states would have the right to establish a 200-mile exclusive resource zone.

This study will first examine the arrangements that presently apply to the Northeast Atlantic fisheries and then review the implications of three possible developments: (1) an extension of the exclusive fisheries jurisdiction of the region's coastal states; (2) the creation of a special body with full authority over the living resources of the Northeast Atlantic; and (3) improvements in the present treaty arrangements.

THE PRESENT TREATY SYSTEM

The present treaty arrangements applicable to the Northeast Atlantic fisheries deal primarily with four issues. These are: (1) the allocation of the area's living resources; (2) the conservation of these resources; (3) fisheries research; and (4) the conduct of fishing operations.[5]

Allocation

A most decisive element in the allocation of the sea's living resources is the breadth of the exclusive fisheries waters of coastal states: fishing in such waters is reserved to the nationals of the coastal state concerned, while high seas fisheries are open to the nationals of all countries. Almost all coastal states of the Northeast Atlantic Ocean have a 12-mile exclusive fishery zone.[6] Iceland is the exception: on September 1, 1972, it established an exclusive fisheries zone of 50 miles[7] —a step that led to a dispute with Great Britain, which was resolved by an agreement between the two governments on November 13, 1973.[8] The nearly universal acceptance of the 12-mile limit has been formalized in the European Fisheries Convention of March 9, 1964.[9] Belgium, Denmark, the Federal Republic of Germany, France, Ireland, the Netherlands, Portugal, Spain, Sweden, and the United Kingdom are parties to this Convention—Norway and Iceland being the exceptions. The convention provides that a coastal state has exclusive jurisdiction over all fisheries within a zone of 6 miles from the baseline of its territorial sea.[10] Within the belt between 6 and 12 miles from the territorial sea, the right to fish may be exercised by the coastal state and by the other parties to the convention that have "habitually" fished in this area between 1953 and 1963.[11] These other states may not substantially alter the nature of their fishing operations,[12] while the coastal state may regulate all fisheries in the outer zone of 6 miles.[13] The parties to the convention also agree to give each other most-favored-nation treatment with respect to the fisheries in the 12-mile zone,[14] but this provision does not prevent the establishment of special regimes between, *inter alia*, the members of the European Economic Community.[15] In conjunction with the European Fisheries Convention a number of agreements were signed that made provision for certain transitional arrangements for the fisheries in the first zone of 6 miles from the coast, but these agreements have expired in the meantime.[16]

These allocation arrangements in the Northeast Atlantic do not constitute a departure from the traditional international law of the sea. However, this cannot be said of a second factor determining the allocation of the region's living resources: the European Economic Community. On October 20, 1970, the EEC adopted two regulations: one concerning the establishment of a common structural policy for the fishing industry and one concerning the common organization of the market in fishery products.[17] Article 2 of the first regulation provides that the member states must ensure equal conditions of access to waters subject to their jurisdiction for all fishing vessels flying the flag of an EEC member state and registered in Community territory. Thus, any exclusive fishery zones established by the EEC member states would have no effect *vis-à-vis* other EEC member states, but only *vis-à-vis* non-EEC members. However, this arrangement was changed by the Treaty of Accession concerning the entry into the Common Market of Denmark, Great Britain, and Ireland. Article 100 of the treaty allows old and new member states to derogate from the requirement of equal access until December 31, 1982, by authorizing them to restrict access to fishing areas within 6 miles from the coast to vessels that have traditionally fished there and that operate from ports situated in the area in question. Article 101 extends the areas that can be reserved to 12 miles for most of the Atlantic coast of the nine member states. The Council of Ministers of the European Communities must decide with respect to the arrangements that will apply after December 31, 1982.[18] The result of these EEC measures is that at least until 1983 the EEC member states may maintain "traditional" exclusive fishery zones of either 12 or 6 miles not only *vis-à-vis* non-EEC member states, but also *vis-à-vis* other EEC member states. However, two non-EEC members—Sweden and Portugal—have special rights to fish within the exclusive fisheries zones of some EEC states as parties to the 1964 European Fisheries Convention; they may continue to carry out traditional fishing operations in the belt between 6 and 12 miles from the coast of the EEC states.

The EEC may soon have another effect on the allocation of the Northeast Atlantic's living resources. It is preparing a regulation which would allow fishermen of one member state to establish themselves freely and without discrimination in the territory of another member state. If this regulation would become effective, Dutch fishermen could, for example, settle in the United Kingdom and catch herring in British territorial waters—something they have been trying to do since the sixteenth century. Then, the EEC would have become in fact a single bloc for purposes of allocation. This would also concern national quota agreements.

A third aspect of allocation in the Northeast Atlantic Ocean concerns the area's high seas fisheries. Generally, there are two methods for allocating high seas fisheries: (1) through free competition among fishing nations; or (2) through a special agreement on national quotas.[19] The first method still prevails in the fisheries of the Northeast Atlantic Ocean, which implies that a nation's share in the catch of a particular fishery is a function of its success in competing for the available catch with the other fishing nations. However, this situation is changing. At its 1970 meeting the North-East Atlantic Fisheries Commission adopted a resolution in which it requested an amendment of Article 7 of the North-East Atlantic Fisheries Convention which would give it the authority to propose: (1) measures for the regulation of the amount of the total catch and its allocation among the member states; and (2) measures for the regulation of the total amount of fishing effort and its allocation among the member states.[20] As this amendment of the convention has not become effective,[21] the commission has requested its members to develop national quota arrangements outside the commission's framework.[22] An example of such an arrangement is the Faroes Fisheries Agreement of September 20, 1973, in which Belgium, Denmark, the Federal Republic of Germany, France, Norway, Poland, and the United Kingdom agreed on national quotas for the high-seas fisheries outside the Faroes exclusive fisheries zone.

Conservation

The conservation of the area's living resources is the concern of the North-East Atlantic Fisheries Convention (NEAFC).[23] This convention established the North-East Atlantic Fisheries Commission,[24] which replaced the Permanent Commission of 1946.[25] NEAFC's present members are: Belgium, Denmark, the Federal Republic of Germany, France, Iceland, Ireland, the Netherlands, Norway, Poland, Portugal, Spain, Sweden, the U.S.S.R., and the United Kingdom. The commission's area of competence comprises roughly the waters between Cape Farewell in the northwest and Novaya Zemlya in the northeast to the latitude of Gibraltar (36° N) in the south; it is subdivided into a number of regions and contiguous with that of the International Commission for the Northwest Atlantic Fisheries. NEAFC consists of two delegates from each member state, who may be accompanied by experts and advisers. It usually meets once a year in London. The commission has a number of regional committees, while its staff consists of a part-time secretary. NEAFC's central responsibility is to "....consider... what measures may be required for the conservation of the fish stocks and for the rational exploitation of the fisheries. ..."[26] To achieve this purpose, the commission may recommend the following measures to its member states: mesh size regulations, fish size regulations, closed areas, closed seasons and gear regulations.[27] Additionally, the commission may request its members to grant it the authority to recommend measures "...for regulating the amount of total catch, or the amount of fishing effort in any period, or any other kinds of measures for the purpose of the conservation of the fish stocks in the Convention area. ..."[28] In 1970 the commission used this provision by requesting the authority to propose measures for regulating the total catch and/or effort and its allocation.[29] Recommendations of NEAFC become effective on a date determined by the commission except for those states that have lodged a formal objection within ninety days after the date of notice of the recommendation in question.[30]

Most conservation measures adopted through NEAFC are concerned with minimum mesh sizes and the minimum size of certain species: of the eleven recommendations in force seven belong to this category. In addition to these measures the commission has established closed area/closed season regulations for high seas salmon fisheries and for North Sea herring fisheries, while it has prohibited trawling in certain areas of the Bay of Biscay and purse-seines in the Celtic Sea. The commission as such may not establish any overall catch limits, but overall quotas have been adopted outside the commission's framework, e.g., for the Atlanto-Scandian herring fisheries.[31] Moreover, NEAFC has indirectly set quotas for the North Sea herring fisheries by allowing each member state to catch a certain tonnage as an exemption from the closed season regulation.

The EEC is also directing its attention to conservation problems. In preparing the resolutions mentioned above,[32] the European commission had originally proposed to adopt a common fisheries policy *vis-à-vis* non-EEC member states. Such a common policy would also involve conservation. However, the Council of Ministers did not accept this proposal: it decided to leave the external fisheries policy in the hands of the individual member states. The commission's idea was reduced to a phrase in the preamble of one of the resolutions noting that "....it must be possible to take Community measures with a view to safeguarding the resources present in the waters in question," i.e., the waters in which the fishermen of the community operate. This theme was taken up by the Treaty of Accession between the EEC and Denmark, Ireland, and the United Kingdom. Article 102 of the treaty states that within six years after the accession of these countries the Council of Ministers must determine "...conditions for fishing with a view to ensuring protection of fishing grounds and conservation of the biological resources of the sea." It seems reasonable to conclude that if this provision will be implemented, the result will be a common policy of the EEC states *vis-à-vis* other states in matters of conservation.

Research

The International Council for the Exploration of the Sea (ICES)[33] is responsible for fisheries research in the Northeast Atlantic Ocean. Although the council was originally established in 1902, its most recent constitution was adopted in 1964.[34] The members are: Belgium, Canada, Denmark, the Federal Republic of Germany, Finland, France, Iceland, Ireland, Italy, the Netherlands, Norway, Poland, Portugal, Spain, Sweden, the United Kingdom, the United States, and the U.S.S.R.; each member state sends two delegates to the council's meetings. Throughout its existence ICES' main function has been to coordinate and encourage the marine research of its member states; only occasionally did it conduct its own investigations. Under the 1964 constitution the council's functions are: (1) to promote research for the study of the sea, particularly with respect to the living resources; (2) to draw up research programs; and (3) to publish the results.[35] ICES is concerned with the Atlantic Ocean and especially with the North Atlantic. The council conducts its business through a number of committees and working groups; presently there are twelve standing committees, each of which deals with a specific area or topic. A consultative committee acts as a coordinating body. In addition to these committees ICES has a bureau with some executive responsibilities and a secretariat.

ICES has been instrumental in encouraging and coordinating the marine research efforts of its members, while it has collected and published a wealth of hydrographic and statistical data; it has also published a wide variety of other studies. An important reason for ICES' effectiveness is its ability to draw practicing scientists into its work. The council also made a substantial contribution to the work of NEAFC. The close cooperation between the two bodies stems, first, from the fact that membership in ICES and NEAFC is almost identical[36] and, second, from the requirement that NEAFC must seek "...when possible..." the advice of the Council.[37] The council has established a special liaison committee to transmit information from ICES to NEAFC and to provide NEAFC with scientific advice. This committee consists of the chairmen of those standing committees of ICES whose work is of interest to the commission. As a result of its relationship with ICES, NEAFC has not established any special research committees of its own.[38]

Conduct

The Northeast Atlantic treaty system is also concerned with the manner in which fishing operations are carried out. The first agreement dealing with this issue was the North Sea Fisheries Convention of 1882, which will be replaced by the Convention on Conduct of Fishing Operations in the North Atlantic of 1967.[39] Like its predecessor, the Conduct Convention sets standards for the marking of fishing vessels and gear, provides for additional signals to be used by fishing vessels and gives rules concerning the operation of vessels. The convention also sets up a system of mutual enforcement. On the high seas the rules laid down in the convention are enforced *vis-à-vis* a fishing vessel not only by the authorized officers of the vessel's flag state, but also by the authorized officers of any state that is a party to the convention.[40] An authorized officer may board a vessel if he has reason to believe that it is not complying with the provisions of the convention. If a breach has occurred, the officer must secure information about the relevant facts and draw up a report concerning the infraction; the vessel's flag state must take further action.

Essentially the same approach to enforcement is followed by the Scheme of Joint Enforcement, which was adopted by the North-East Atlantic Fisheries Commission at its fifth annual meeting.[41] The scheme provides that the fishery inspectors of all NEAFC member states may observe whether or not the vessels of any member state comply with NEAFC's conservation measures. If they establish that an infraction has occurred, inspectors submit a report to their flag state, while copies are being sent to the vessel's flag state and to the commission. Although the Scheme of Joint Enforcement had a difficult start, most NEAFC member states participate now in its implementation. Countries that had refused to accept inspection on and/or below deck are gradually withdrawing their objections.[42]

AN EVALUATION OF THE PRESENT
TREATY SYSTEM

It is a positive aspect of the present allocation arrangements in the Northeast Atlantic Ocean that there is relatively little disagreement concerning the extent of a coastal state's exclusive fisheries jurisdiction. However, the treaty system has been unable to accommodate the special interests of Iceland, and the extent of Iceland's fisheries zone continues to be an object of international disputes.

Another unsatisfactory aspect of the Northeast Atlantic's allocation arrangements is the fact that most high seas fisheries are still being allocated on the basis of free competition. Allocation through free competition results in a great deal of friction among the nations and fishermen participating in a particular fishery. It also forces them to apply extra fishing effort for the sole purpose of being "competitive." Consequently, free competition contributes to the overcapitalization of the fishery. In view of the urgency of these problems it is unacceptable that NEAFC has not yet the authority to make recommendations on national quotas. However, this should not prevent the commission from trying to reach agreement as soon as possible on the *general* criteria for the national quota schemes it might propose in the future. In this respect it is fortunate that in NEAFC the confrontation between coastal fishing countries and distant water fishing countries is not as intense as in the International Commission for the Northwest Atlantic Fisheries. In the Northeast Atlantic Ocean another confrontation may become more important, i.e., between the member states of the EEC[43] and the non-EEC states.[44] Much will depend upon the pace at which the EEC states merge into a single bloc for purposes of allocation.

Although NEAFC has been unable to prevent the overfishing of such important stocks as North Sea herring, Atlanto-Scandian herring, and cod, there is little doubt that the commission's conservation programs have contributed to the consistently high catches of the Northeast Atlantic fisheries. In this respect the commission was helped by the fact that in the past the expanding fishing effort could be directed at unused or underutilized species. However, virtually all commercially attractive stocks are now heavily exploited, which implies that there is little opportunity for absorbing additional effort. This, and the fact that in all probability fishing effort in the Northeast Atlantic will continue to expand,[45] means that the ultimate test of NEAFC's effectiveness in respect of conservation still lies in the future. Therefore, it is imperative that the commission obtain as soon as possible the authority to recommend total allowable catch limits. It would also be desirable if NEAFC's conservation programs were reinforced by those of the EEC. However, it appears unlikely that the EEC will agree on a common conservation policy in the near future.

Arrangements for fisheries research are quite satisfactory insofar as the biological aspects are concerned. ICES ensures both the quality and the impartiality of the biological information available to NEAFC. However, the commission lacks advice concerning the economic, social, and legal aspects of its regulatory work. This is unacceptable if the commission must deal with allocation and overcapitalization problems, since these are primarily economic in nature. Therefore, the scope of fisheries research in the Northeast Atlantic must be expanded. More generous funding would also improve the quantity and quality of the available data base.

The central question with respect to the conduct of fishing operations no longer concerns the legal arrangements, but rather the willingness of states to participate in their implementation. No legal arrangements concerning enforcement can be effective without an adequate number of inspectors at sea or in port.

The present Northeast Atlantic treaty system has not dealt with the overcapitalization of the region's fisheries. NEAFC has only discussed the regulation of fishing effort in very general terms, even though it is responsible not only for the conservation of the region's fish stocks, but also for a "....rational exploitation of the fisheries"[46]—a phrase which appears to bring the economic status of the fisheries under the commission's mandate. However, the commission has requested the authority to make recommendations concerning total effort limits and concerning the allocation of fishing effort among fishing nations. If accepted, this would enable NEAFC to regulate fishing effort and to deal directly with overcapitalization problems. An indirect step toward solving these problems would be taken if the commission were to make recommendations on national quotas. National quotas determine each nation's share in a particular fishery in advance, and thus eliminate to a large extent the need for competition. This enables each nation to decide for itself how much effort it wishes to apply. Some states would undoubtedly reduce fishing effort, but there could be countries that would consider the protection of employment opportunities more important; they would be free not to reduce effort. Consequently, national quotas allow each state to deal with overcapitalization in accordance with its own priorities. This is another reason why it is imperative that NEAFC obtain as soon as possible the authority to formulate national quota schemes.

Alternative Regimes

The Northeast Atlantic treaty system could evolve in a number of directions: national fisheries limits could be extended, a central authority could be set up, preferential rights schemes could be introduced, existing arrangements could be improved, etc. If carried to its extreme, each option would represent a distinct alternative to the present system. However, it is unlikely that this will occur: in reality the future Northeast Atlantic treaty system will emerge from a combination of developments even though such developments may have their origin in theoretically incompatible options. This implies that it is a simplification to review the alternatives to the present treaty regime individually. Nevertheless, the following sections will adopt this simplification for the sake of clarity.

An Extension of Exclusive Fishery Limits

A first alternative to the present Northeast Atlantic treaty system would involve an extension of the exclusive fisheries jurisdiction of the region's coastal states.[47] The effect of such a step would depend upon the distances and the number of states concerned, but it can be best discussed by assuming that each coastal state of the area would establish an economic resource zone of 200 miles.[48] Most of the Northeast Atlantic Ocean and virtually all commercially attractive fishing grounds would be covered by such a 200-mile limit. The

North Sea area would be completely divided up among the coastal states, while the same would occur in the Irish Sea and the British Channel.

Establishing economic resource zones of 200 miles in the Northeast Atlantic would radically alter the allocation of the area's living resources. States with long coastlines facing the open ocean—Portugal, Ireland, Norway, and Iceland[49]—would clearly benefit, but the German Democratic Republic, Poland, and the U.S.S.R. would definitely lose: they fish in the area, but have no—or only a very short—coastline bordering the Northeast Atlantic Ocean. However, for most countries the effect would be more ambiguous. Their geographical location may prevent them from claiming the full breadth of 200 miles,[50] or any gains from the creation of an economic resource zone may be offset by losses in the distant water fisheries. In the Northeast Atlantic Ocean no clear distinction exists between coastal fishing nations and distant water fishing nations. As a result, it is difficult for countries like France, the Federal Republic of Germany, the United Kingdom, Denmark, and the Netherlands to calculate whether, from the viewpoint of allocation, establishing an economic resource zone will result in a net gain or in a net loss. However, the fact that in the U.N. Seabed Committee almost all Northeast Atlantic coastal states[51] have refrained from supporting an extension of a coastal state's exclusive fisheries jurisdiction appears to reflect not only tradition, but is in all probability also indicative of their uncertainty as to its effect.

An extension of the exclusive fisheries jurisdiction of coastal states could be a remedy for the ineffectiveness of the present international conservation arrangements. Conservation measures would no longer depend upon the difficulties involved in reaching international agreement among all states fishing in the Northeast Atlantic, but could be adopted through the coercive powers of the region's coastal states.[52] The adoption of effective conservation measures would be in the self-interest of a coastal state once it has exclusive jurisdiction over the resources concerned. However, many species of fish would continue to migrate across the boundaries of a coastal state's economic resource zone; some herring stocks, for example, are known to migrate in a circlelike pattern in the North Sea. This could make it impossible for a single coastal state to adopt effective conservation measures, which suggests that in these cases international cooperation would remain a necessity.[53] Moreover, in many situations in which coastal states could have acted by themselves, they have been as ineffective as international arrangements in adopting conservation measures. The fate of the Atlanto-Scandian herring stocks off the Norwegian coast offers here a striking illustration. The conclusion must be that 200-mile economic resource zones would not necessarily lead to more effective action in respect of the conservation of the Northeast Atlantic's living resources.

Establishing 200-mile economic resource zones would be a major step toward solving the overcapitalization problems of the Northeast Atlantic's fisheries: as in such a zone a coastal state would have exclusive jurisdiction, it could control fishing effort. Coastal states could use this authority to limit fishing effort to a minimum and to eliminate any overcapitalization; a coastal state could also apply more effort than strictly necessary if it would wish to protect employment opportunities. However, by eliminating the freedom of fishing, economic resource zones create the conditions for dealing with over-capitalization problems.

Fisheries research is inherently an international activity: fish migrate regardless of manmade boundary lines and their life patterns may be influenced by factors originating outside the areas of exclusive coastal state jurisdiction. Consequently, it would be undesirable for coastal states to use their exclusive jurisdiction over an economic resource zone for the purpose of imposing unreasonable limitations on fisheries research. However, it appears unlikely that this would happen in the Northeast Atlantic: the states of this region have a strong tradition of international cooperation in respect to marine research. Therefore, establishing economic resource zones would in all proba-bility have no serious adverse effects on the area's research.

Many of the above observations regarding the effects of economic resource zones apply not only to the Northeast Atlantic, but to other areas of the seas as well. This is not true for two additional questions raised by the creation of such zones in the Northeast Atlantic Ocean: (1) are they in accord-ance with the 1964 European Fisheries Convention; and (2) are they in accord-ance with the EEC Regulations mentioned above?[54] The 1964 European Fisheries Convention is of "unlimited duration," while the convention may not be denounced within twenty years from the entry into force, which occurred in 1966. As the regime of the convention is clearly incompatible with a 200-mile economic resource zone, it must be concluded that until 1986 a party to the convention may establish such a zone only with the consent of all other con-tracting states. Norway and Iceland are the only Northeast Atlantic coastal states that are not subject to this condition; they are not parties to the 1964 European Fisheries Convention. As far as the second question is concerned, it has been mentioned that until 1983 the EEC members may derogate from the requirement of equal access to national fishery waters and may reserve certain fishing areas for their own fleets.[55] However, these areas may not be located beyond 12 miles from the coast. Accordingly, it would be a violation of the applicable EEC rules if a member state would refuse the fishermen of the other EEC states access to its 200-mile economic resource zone. Consequently, no member of the EEC may establish an economic resource zone in which fishing is reserved for its own nationals without the consent of the other member states.

These observations lead to the conclusion that 200-mile economic resource zones resolve few problems of the Northeast Atlantic's fisheries.[56] Moreover, in the Northeast Atlantic Ocean there are two groups of states that can establish truly exclusive 200-mile zones only by mutual agreement: the parties to the 1964 European Fisheries Convention and the members of the EEC. This may suggest that it is unlikely that 200-mile economic resource zones will become generally accepted in the Northeast Atlantic in the near future. However, there can be little doubt that if the 1974 United Nations Conference on the Law of the Sea decides that all coastal states are entitled to a 200-mile economic resource zone, most Northeast Atlantic coastal states will also establish such a zone. It would be impossible to uphold in the Northeast Atlantic the freedom of fishing on the high seas if most other fishing areas of the seas would be subject to exclusive coastal state jurisdiction.

A Central Authority

A second possibility would be to establish a special Northeast Atlantic fisheries body with full and exclusive authority over all living resources in the area.[57] Presumably, this body could be NEAFC, but—on a smaller scale and with more limited participation—it could also be the EEC. In the latter approach all EEC member states could, for example, first establish economic resource zones of 200 miles and then agree to vest the authority over the resources of these zones in an EEC organ. However, it will be assumed here that full authority over the living resources of the Northeast Atlantic Ocean will be vested in an organization covering the whole area, e.g., NEAFC.

Such a step could have several advantages. Its exclusive jurisdiction over the stocks would enable the organization to adopt and enforce all necessary conservation measures; in this respect its forcible powers could be similar to those of coastal states; The organization could also deal with overcapitalization problems. It could, for example, operate the Northeast Atlantic fisheries itself and buy the necessary manpower and equipment in the cheapest market, while selling the fishery products in the dearest market; the organization's profits could be redistributed among its member states. Alternatively, the organization could auction off rights to fish in the Northeast Atlantic to the highest bidders, which would also ensure the economic efficiency of the fisheries. It could also deal with overcapitalization problems by restricting fishing effort through a system of licenses issued either to states or directly to fishing companies.[58] The fact that all member states could share in the revenues derived by the organization from the Northeast Atlantic's fisheries would also constitute a major step toward solving the area's allocation problems. If, for example, the organization adopted the bidding system, states that bid successfully would share in the wealth of the Northeast Atlantic's living resources by carrying out the actual fishing operations, while states that did not bid successfully would still acquire some of the resources' wealth by sharing in the organization's revenues. Finally, the organization could reserve part of its income for conducting all necessary scientific investigations and thus solve the problems of fisheries research.

However, several disadvantages would outweigh these advantages.

First, operating on the basis of economic rationality, the organization would in all probability limit fishing effort to the level at which it would maximize profits, rather than to the level at which it would maximize the catch. The resulting reduction of the catch of the Northeast Atlantic's fisheries could be unacceptable, in view of a worldwide shortage of animal protein. Second, elimination of overcapitalization problems might not be a first priority of all states fishing in the Northeast Atlantic; some may prefer to protect employment opportunities instead. This would undermine one of the most important justifications for establishing an organization with full authority, i.e., the fact that it could ensure the economic efficiency of the fisheries. Third, if the organization were to buy manpower and equipment in the cheapest markets and sell the fishery products in the dearest markets, the result could be that the poorest nations—or part of nations—of the Northeast Atlantic would carry out all fishing operations, while only the richest nations—or parts of nations—could afford to buy the fish products, which appears to be an unacceptable allocation of the region's living resources. Fourth, serious difficulties would in all probability arise in connection with the distribution of the organization's profits. Should only coastal states share in these profits or also states that have traditionally fished in the region? If so, in what proportion? Should there be a share for the world community as a whole? These questions could very well lead to serious friction among the member states, which could frustrate the organization's functioning. This already suggests the most fundamental obstacle to the creation of an organization with full authority over the Northeast Atlantic's living resources: it would be such a radical departure from existing structures and patterns that the states of the region would be unwilling to accept it. The establishment of such an organization must be deemed beyond present political realities. It appears that only a major disaster in the fisheries could alter this conclusion.

Improving Existing Arrangements

The above two possibilities represent the extremes of the continuum: one vests the authority over the living resources of the Northeast Atlantic in the coastal states, while the other vests such authority in an international organization. A third option is less extreme: to improve the existing arrangements.[59] The shortcomings of the present Northeast Atlantic treaty system have been reviewed above.[60] The general conclusion was that it is imperative that NEAFC obtain as soon as possible the powers it requested under Article 7 of the North-East Atlantic Fisheries Convention. These powers would allow the Commission to deal with allocation problems by recommending national quotas, with questions of conservation by recommending overall catch limits and with the overcapitalization of the fisheries by recommending either direct effort controls or national quotas. However, without a political willingness of all states concerned to accept improvements in the present treaty system, NEAFC will not obtain the necessary powers and, even if it did get these powers, it would not be able to use them effectively. Therefore, the crucial question is how the essential condition of political willingness can be met. It appears that there are two alternatives: (1) through a disaster in all major fisheries of the Northeast Atlantic Ocean—similar to the collapse of, for example, the Atlanto-Scandian herring fisheries; or (2) by providing the states in question with an incentive to accept an improved management system. Ignoring the first alternative, it seems that the states fishing in the Northeast Atlantic would have a stronger interest in effective management if two more basic changes were made in the present Northeast Atlantic treaty system. These are: (1) the states fishing in this area should have collectively an exclusive right of access to the fisheries; and (2) the area's coastal states should have certain preferential rights over the stocks off their coasts.

The freedom of fishing on the high seas has become in many respects an impediment to effective management. It implies that: (1) all states have a right of free and unrestricted access to any high seas fisheries; and (2) the only way in which states can share in the wealth of the living resources of the high seas is by actually carrying out fishing operations. Free and unrestricted access means that outsider-states can reap the benefits of any conservation measures adopted by the states participating in a particular fishery. As a result, it is unattractive for these states to accept conservation measures: they have no guarantee that they will be the only ones to benefit if a stock recovers as a result of their restraint. Free and unrestricted access is also a direct cause of overcapitalization problems since it is incompatible with controlling the effort applied to a certain fishery; it forces states to engage in wasteful competition. Moreover, it is difficult for states participating in a certain fishery to accept a national quota scheme which limits their catch to a very distinct level as long as outsider-states can freely enter the fisheries to which such a scheme applies. Although some national quota arrangements attempt to deal with this problem by reserving a certain percentage of the total allowable catch for outsiders, the fact remains that the problem of new entrants remains a most serious obstacle to agreement on national quotas.

The second consequence of the freedom of fishing has its impact particularly on allocation problems. The fact that a state can share in the wealth of the living resources of the high seas only by actually carrying out fishing operations implies that it is unattractive for a state to give up its right to enter any high seas fishery, even though in fact it may never exercise this right. If, on the other hand, a state could share in the wealth of high seas stocks without actually fishing, it would be relatively easier to give up such rights of access, which could help solve the problem of outsider-states mentioned above. The conclusion must be that the freedom of fishing on the high seas has become a major reason for the shortcomings of many international fisheries regimes. The Northeast Atlantic Ocean treaty system is no exception.

An alternative to the freedom of fishing would be to restrict access

to the fisheries of the Northeast Atlantic Ocean to those states that had partici-
pated in these fisheries for some time. Only these states would have a right of
access, whereas other states would not be allowed to enter a Northeast Atlantic
fishery except with the consent of the states with a right of access. The result
would be that in each fishery a specific number of states could participate and
that these states could deal with conservation, allocation, and overcapitalization
problems without the threat of new entrants. Thus, as a group these states
would have an exclusive right to the stock exploited by the fishery in question,
which implies that rational regulation would be in their self-interest. Consequently, closing access to the Northeast Atlantic's fisheries would be a major
step in bringing about the political willingness required for more effective fisheries management. The interest of the states that would be excluded from these
fisheries could be protected by imposing certain levies on states with a right of
access and by redistributing the revenues to states without a right of access. Such
an arrangement would allow states without a right of access to share in the
wealth of the living resources of the Northeast Atlantic, even though they could
no longer engage in fishing operations in that area. Levies would not necessarily
increase the price of fishery products, since they could be paid from the gains
achieved by the economic rationalization of the fisheries made possible by the
closure of access.

    Closure of access does not necessarily constitute a totally new
approach. For example, its practical effect could be very similar to that of a
decision by all Northeast Atlantic coastal states to establish a *common* 200-
mile economic resource zone. The major difference appears to be that closing
access on the basis of past performance automatically recognizes the rights of
distant-water fishing countries, while in the common economic zone approach
such rights would have to be negotiated.

    The second mechanism for enhancing the effectiveness of fisheries
management in the Northeast Atlantic would involve the adoption of certain
preferential rights for the region's coastal states. It is a well-documented fact
that some countries, e.g., Iceland and Norway, are particularly dependent upon
marine fisheries, while in other countries, e.g., the Faroes and the United Kingdom, this is true for specific regions. A continued refusal to acknowledge this
dependency will lead to a great deal of friction among the Northeast Atlantic
states, which will undermine the effectiveness of the area's management arrangements. The recent dispute between Iceland and the United Kingdom has been,
for example, the most important reason why Iceland has not yet accepted the
granting of additional powers to NEAFC. Thus, a dispute arising from a refusal
to recognize a country's special dependency on marine fisheries has had a negative impact on the functioning of NEAFC.[61] Coastal state preferential rights
could accommodate the various interests involved in such disputes. They could
give recognition to the special dependency of certain coastal populations on
marine fisheries with a minimum of damage to the interests of distant water
fishing states. Preferential rights could involve for a coastal state the right to

exploit exclusively certain high seas fishing grounds or stocks, the right to use a particular type of gear that noncoastal fishermen would not be allowed to use, a higher share in a national quota arrangement, etc. The recently concluded Faroes Fisheries Agreement is an example of a preferential-rights arrangement: it protects the catch of the Faroese fishermen, while reducing the catches of the distant-water fishermen. However, coastal state preferential rights should become more generally accepted in the Northeast Atlantic Ocean,[62] otherwise, the states of that region might become involved in a series of disputes that would jeopardize any steps to improve the present management arrangements. NEAFC could make a contribution towards the wider acceptance of coastal state preferential rights by including them, if necessary, in the national quota arrangements which the commission may propose under the amended Article 7. However, it is doubtful that in the Northeast Atlantic Ocean coastal state preferential rights could be similar to the arrangements that have been proposed in the U.N. Seabed Committee.[63] Essentially, these U.N. proposals would give coastal states the right to reserve for their own fishermen such portion of the allowable catch of the living resources off their coasts as these fishermen can harvest. As virtually all coastal states of the area have an extensive fishing capacity, the application of this principle to the Northeast Atlantic would mean that these coastal states would harvest all offshore resources and that there would be little room for distant-water fisheries. Consequently, basing coastal state preferential rights on fishing capacity implies for the Northeast Atlantic that in fact coastal state preferential rights would be indistinguishable from coastal state economic resource zones.

FINAL REMARK

Whatever the future regime of the Northeast Atlantic's fisheries will be, it seems reasonably safe to suggest that it will incorporate many aspects of the present treaty arrangements. This would be the case even if the area's coastal states were to establish 200-mile economic resource zones—such zones, for example, alter the context within which NEAFC functions, but do not undermine the need for its existence. This implies that it remains desirable to improve the existing fisheries treaty system in the Northeast Atlantic, even though there is uncertainty as to the outcome of the current developments in the law of the sea generally.

Notes to Chapter 4

1. 15.4 million tons and 13.7 million tons respectively.
2. FAO, *Yearbook of Fishery Statistics 1970* (Rome 1971), p. 436.
3. E.g., blue whiting in the Norwegian Sea and squids.

4. XI *International Legal Materials* (1972), p. 643.

5. For a more detailed discussion of these problem areas, see A. W. Koers, *International Regulation of Marine Fisheries. A Study of Regional Fisheries Organizations* (London; Fishing Books Ltd., 1973), pp. 39–77.

6. Belgium, Denmark, France, Ireland, Norway, Portugal, Spain, Sweden, and the United Kingdom; the Federal Republic of Germany and the Netherlands have rights to such a zone, but have not established it.

7. *Supra*, note 2.

8. The British catch around Iceland has been limited to 130,000 tons, while the trawler fleet may not have more than 139 vessels—freezer trawlers and factory ships are prohibited altogether.

9. III *International Legal Materials* (1964), p. 476.

10. Art. 2.

11. Art. 3.

12. Art. 4.

13. Art. 5.

14. Art. 8.

15. Art. 10.

16. III *International Legal Materials* (1964), pp. 491, 493.

17. Respectively, Regulation (EEC) No. 2141/70 and Regulation (EEC) No. 2142/70.

18. Art. 103.

19. S. Oda, "Distribution of Fish Resources of the High Seas: Free Competition or Artificial Quotas?", in *The Future of the Sea's Resources*, ed. L. M. Alexander, Law of the Sea Institute 2nd Conference, 1967 (Kingston, R.I., 1968), pp. 29–32; J. A. Crutchfield, "*National Quotas for the North Atlantic Fisheries: an Exercise in Second Best,*" in *International Rules and Organization for the Sea*, ed. L. M. Alexander, Law of the Sea Institute 3rd Conference, 1968 (Kingston, R.I., 1969) pp. 263–75.

20. NEAFC, *Report of the Eighth Meeting*, London, 1970, p. 12.

21. Iceland and Belgium have not yet accepted it.

22. NEAFC, *Report of the Eighth Meeting*, London, 1970, p. 11.

23. London, January 24, 1959, 486 U.N.T.S. 157.

24. Hereinafter referred to as NEAFC.

25. 231 U.N.T.S. 200.

26. Art. 6, para 1.

27. Art. 7, para. 1.

28. Art. 7, para. 2.

29. *Supra*, p. 5.

30. Art. 8 and art. 9.

31. NEAFC, *Report of the Ninth Meeting*, London, 1973, p. 19.

32. *Supra*, p. 3.

33. Hereinafter referred to as ICES.

34. III *International Legal Materials* (1964), p. 302.

35. Art. 1.

36. Finland and Italy are members of ICES, not of NEAFC.

37. Art. 11, North-East Atlantic Fisheries Convention.

38. In this respect it differs from the International Commission for the Northwest Atlantic Fisheries, which has its own scientific advisory bodies.

39. VI *International Legal Materials* (1967), p. 760.

40. Art. 9.

41. NEAFC, *Report of the Fifth Meeting*, London, 1968, p. 79.

42. NEAFC, *Report of the Tenth Meeting*, London, 1972, p. 11.

43. Belgium, Denmark, the Federal Republic of Germany, France, Ireland, the Netherlands, and the United Kingdom.

44. Iceland, Norway, Poland, Portugal, Spain, Sweden, and the U.S.S.R.

45. ICNAF has imposed rather stringent limitations on some distant water fisheries of the U.S.S.R., Poland, the German Democratic Republic, and the Federal Republic of Germany and it must be expected that these countries will divert some of their surplus effort to the Northeast Atlantic.

46. Art. 6, para. 1 (b) North-East Atlantic Fisheries Convention.

47. Koers, *op. cit.*, note 3, pp. 233–43.

48. This would be in line with several proposals submitted to the U.N. Seabed Committee, e.g., Malta, *Draft Ocean Space Treaty*, A/AC. 138/53, August 23, 1971; Kenya, *Draft Articles on Exclusive Economic Zone Concept*, A/AC/138/SC. II/L.10, August 7, 1972; Australia and New Zealand, *Principles for a Fisheries Regime*, A/AC.138/SC.II/L.11, August 11, 1972; Iceland, *Jurisdiction of Coastal States over Natural Resources of the Area Adjacent to their Territorial Sea*, A/AC.138/SC.II/L.23, April 23, 1973; Canada, India, Kenya, Madagaskar, Senegal, and Sri Lanka, *Draft Articles on Fisheries*, A/AC.138/SCII/L.38, July 16, 1973; Australia and Norway, *Certain Basic Principles on an Economic Zone and on Delimitation*, A/AC.138/SC.II/L.36, July 16, 1973; Ecuador, Panama, and Peru, *Draft Articles on Fisheries in National and International Ocean Space*, A/AC.138/SC.II/L.54, August 10, 1973; and Zaire, *Draft Articles on Fishing*, A/AC.138/SC.II/L.60, August 17, 1973.

49. In the U.N. Seabed Committee Iceland and Norway have supported the general idea of a coastal state economic resource zone, note 37.

50. Belgium, the Netherlands, the United Kingdom, the Federal Republic of Germany, Denmark, and Sweden.

51. With the exception of Iceland and Norway.

52. H. Kasahara, "International Arrangements for Fisheries," in *The United Nations and Ocean Management*, ed. L. M. Alexander, Law of the Sea Institute 5th Conference, 1970 (Kingston, R.I., 1971), p. 39.

53. W. Chapman, "Fishery resources in offshore waters," in *Offshore*

Boundaries and Zones, ed. L. M. Alexander, Law of the Sea Institute 1st Conference, 1966 (Columbus, Ohio, 1967), p. 96.

    54. Art. 15.

    55. *Supra*, p. 3.

    56. In essence, it only establishes conditions for dealing with over-capitalization problems.

    57. Koers, *op. cit.*, note 3, pp. 252-58.

    58. F. T. Christy, Jr., "The Distribution of the Sea's Wealth in Fisheries," in *Offshore Boundaries and Zones*, ed. L. M. Alexander, Law of the Sea Institute 1st Conference, 1966 (Columbus, Ohio, 1967), pp. 116-17.

    59. Koers, *op. cit.*, note 3, pp. 119-28 *passim*.

    60. *Supra*, pp. 9-11.

    61. In this respect the agreement between the two countries came too late, even though it established certain preferential rights for Iceland by reducing the British catch.

    62. This is also recommended by the EEC Commission.

    63. U.S.S.R., *Draft Article on Fishing*, A/AC.138/SC.II/L.6, July 18, 1972; Canada, *Management of the Living Resources of the Sea*, A/AC.138/SC.II/L.8, July 27, 1972; United States, *Revised Draft Fisheries Article*, A/AC.138/SC.II/L9, August 4, 1972; and Japan, *Proposals for a Regime of Fisheries on the High Seas*, A/AC.138/SC.II/L.12, August 14, 1972.

Chapter Five

Biological Consequences of Alternative Regimes

J. L. McHugh

INTRODUCTION

Most fishery disputes rage around two separate and distinct issues: management of the fishery and the resource for maximum (or "optimum") sustainable yield, and who gets the catch. Solution of fishery problems probably would be easier if the scientific questions related to conservation and the social-political problems related to allocation of the resource among users were recognized as distinct, and priorities were established accordingly. But too often social-political issues take precedence over scientific matters, or the two sets of issues are thoroughly entangled, sometimes on purpose. Indeed, the social and the scientific aspects are not always unrelated, for in the absence of agreement on allocation, a rational conservation scheme may be difficult or impossible to establish.

The dilemma is equally troublesome whether the problem be international or entirely domestic. Conflicts between conservation and allocation occupy much of the attention of such diverse international bodies as the

Parts of the analysis on which this paper is based were made under support of a fellowship with the Woodrow Wilson International Center for Scholars, Washington, D.C., July and August 1971.

Information on the fisheries of New York State is a result of research sponsored by NOAA Office of Sea Grant, Department of Commerce, under grant No. 04-3-158-39. The U.S. government is authorized to produce and distribute reprints for governmental purposes notwithstanding any copyright notation that may appear hereon.

71

International Commission for the Northwest Atlantic Fisheries (ICNAF), the
International North Pacific Fisheries Commission (INPFC), the Inter-American
Tropical Tuna Commission (IATTC), and the International Whaling Commission
(IWC). The problem takes different forms according to the special interest of
the affected nation or segment of the fishery. In the Northwest Atlantic, from
the viewpoint of American fishermen or even the residents of coastal New
England as a group, the problem is "the Russians" and the popular solution is
to declare sovereignty over waters out to 200 miles off the coast. In the Northeast Pacific the problem and the solution are seen as equally simple, although
the villain is Japan as well as the Soviet Union. In the eastern tropical Pacific
the roles are reversed. There, frustrated Ecuadorean and Peruvian fishermen
stand on the beach, shaking their fists at the American tuna fleet. With respect
to whaling the issue has become polarized beyond the point of compromise.
Public opinion in the United States, Canada, the United Kingdom and other
parts of the world has been stirred to high emotional levels by relatively small
but effective groups of people who believe with all sincerity that killing of whales
is a moral issue, and that whaling should be brought completely to a halt.

When the issue is purely domestic, in the sense that the problem is
confined to coastal fishery resources that do not traverse waters beyond national
jurisdiction, at least in significant numbers, most people think the problems and
their solutions are equally as simple. A good example is the controversy in the
United States between recreational and commercial striped bass (*Morone
saxatilis*) fishermen. In New York State, as elsewhere, recreational fishermen
have a simple solution: prohibit commercial fishing for the species. They do
not stop at this, but also propose that striped bass caught by commercial fishermen in other states be barred from sale in New York State. The stated justification is conservation of the resource, as it is in the international disputes cited.
The overriding incentive, however, is unilateral access to the resource. Similar
domestic controversies based on allocation of the resource are common between
segments of the commercial fishing industry, and between fishermen and others
who find fishing incompatible with their own use of the coastal environment.

OBJECTIVES OF MANAGEMENT

Conservation or wise management is the implicit objective of fishery laws and
regulations in most countries, including the United States. The objective is
stated explicitly in the nine international fishery conventions to which the
United States is party. The language varies, but not in its essential details, from
the language in the Preamble to the International Convention for the Conservation of Atlantic Tunas (ICCAT):

> The Governments...considering their mutual interest in the popula-
> tions of tuna and tuna-like fishes found in the Atlantic Ocean, and

> desiring to co-operate in maintaining the populations of these fishes
> at levels which will permit the maximum sustainable catch for food
> and other purposes, resolve to conclude a Convention for the con-
> servation of the resources....

and in Article IV:

> ...to carry out the objectives of this Convention the Commission
> shall be responsible for the study of the populations of tuna and
> tuna-like fishes.... Such study shall include research on the abun-
> dance, biometry and ecology of the fishes; the oceanography of
> their environment; and the effects of natural and human factors
> upon their abundance.

In brief, ICCAT and most other fishery conventions propose to manage certain
living resources and their fisheries for maximum sustainable yield, and to conduct broad programs of scientific research to provide the information necessary
to obtain this objective. Presumably, this also is (or should be) the objective
of domestic fishery management programs.

Thus, as far as the stated objectives for management of the resource
are concerned, the five commissions cited are essentially in agreement. Each is
different in the way it deals with the question of allocation of the catch. The
convention which established ICNAF has nothing whatever to say about who
will take the catch, or in what proportion. INPFC, on the contrary, was created
by a convention with closed membership, under which Japan agreed not to fish
for certain species fully utilized by Canada and the United States, even in
certain international waters, as provided by the Principle of Abstention. The
Pacific and Atlantic tuna conventions and the whaling convention contains no
provisions for dividing the catch, and in fact the whaling convention of 1966
specifically prohibits such division. Allocation of the tuna catch is done
through intergovernmental discussions between member nations and nonmem-
bers that fish tunas in the eastern tropical Pacific. The whale catch is divided
by ocean area through negotiation between governments that operate in each
area, after the commission has decided on global quotas. In ICNAF, IATTC,
ICCAT, and IWC the problem of allocation was not cause for concern when the
conventions were negotiated. In the first two the need for allocation arose
when establishment of quotas or substantial increases in fishing effort led to
intense competition between fleets to take the major share of the allowable
catch. Such scrambles, aptly named the "olympic system" by the Japanese
when Antarctic whaling quotas led to building of bigger, more efficient catcher
boats and factory ships at great cost, soon bring the competing nations to the
negotiation table, for economic reasons alone. The bitterness and difficulty of
the arguments reached a peak in the eastern tropical Pacific when the growing

interest of other nations in the tuna resources brought demands for ever-increasing shares of the quota.

In the North Pacific the work of INPFC has also increased in difficulty with time as continued heavy fishing pressure by Japan has added to the number of resources of interest to the commission, and incidental catches of some species, like halibut, have created problems for other management programs. Extension of fishing by the Soviet Union, a nonmember of INPFC, to the continental shelf of the west coast of North America has raised the same kinds of problem, which have had to be discussed bilaterally. Bilateral negotiations also have been resorted to in the Northwest Atlantic when member nations of ICNAF have extended their fishing activities to areas outside the commission regulatory area.

Thus, the evolution of attempts at management of international fisheries in which the United States has an interest usually has been first to set up an international administrative body; then research, to provide a basis for management measures; then regulation of the catch, usually by quota, but also by a variety of other methods; then bilateral or multilateral negotiation to settle special problems; then surveillance and monitoring arrangements; then country quotas; then attempts to persuade or force other nations to refrain from fishing off the coast of the coastal state. The first step by international agreement to broaden the zone of fishery jurisdiction of coastal states was the general agreement at the last Law of the Sea Conference to add a contiguous zone to extend not beyond 12 miles from the baseline used to measure the breadth of the Territorial Sea. The second step has been a series of unilateral declarations establishing broader zones of national jurisdiction, such as the 200-mile declarations by several South American countries. Prevailing opinion in the United States now favors unilateral extension of jurisdiction over fisheries.

Prevailing opinion in the United States also favors the view that international arrangements for management of fisheries have been ineffective. This is why most people in the fishing industry and many others with an interest in fishery resources think that an immediate extension of jurisdiction is necessary. Yet international fishery arrangements, although they have moved too slowly and are far from perfect, have by no means been ineffective. Even IWC and ICNAF, despite overwhelming opinion to the contrary, are doing remarkably well in the face of what often have seemed to be overwhelming odds.

DOMESTIC FISHERY MANAGEMENT

A move to extend national jurisdiction over fisheries to a broader zone off the shores of a coastal state, whatever form it might take, and whether it is implemented by unilateral action or international agreement, will carry with it the obligation to husband the living resources wisely. If the action is unilateral the obligation may be nothing more than a moral obligation. If the action is by international agreement the obligation almost certainly will be an explicit requirement. It is appropriate therefore to examine past domestic performance, to determine how well such obligations have been discharged up to now. It is not enough to show that a resource is in satisfactory condition and to conclude therefore that management has been successful. In addition it must be demonstrated that the management regime has taken positive action to maintain that resource in such a condition that it can yield the maximum sustainable return, or if the resource has been overfished that it is being restored to that condition.

As far as I am aware, such a study has not been made worldwide, and someone ought to undertake it. For the United States I am willing to hazard the assertion that no major coastal fishery is now, or ever has been, managed successfully. The only possible exception that might be offered in rebuttal is that some major Pacific salmon resources are being managed successfully, but this is not an exception to the assertion. Management of some salmon stocks in territorial waters along the west coast of North America came about because it became mandatory through international agreements.

In the United States, although we have no dearth of fishery laws and regulations, and spend considerable amounts of money on fishery research, development, and management, the results have not been impressive. No detailed and comprehensive review of the history of management of United States fisheries has been made, but such a study probably would show much the same general pattern as the history of the coastal fisheries of New York State.[1] The full implications of the New York study become clear only when the history of landings is considered species by species. It is revealing, however, to examine the record by three major commercial categories; food finfisheries, shellfisheries, and industrial fisheries. Landings of food finfish in New York reached a peak of about 25,000 metric tons in 1939, and have been trending downward ever since, to less than half the 1939 maximum in the early 1970s (Figure 5-1). Shellfish landings were at their peak in 1950 at about 11,500 metric tons, dropped to a low of less than 5,000 metric tons in 1958, and have subsequently increased rather steadily to a level of about 7,000 metric tons (Figure 5-2). The industrial fisheries have had two periods of high production, from 1880 to the early 1920s, when the New England menhaden fishery was at its maximum, and from the mid-1940s to mid-1960s. The peak of the modern industrial fishery in New York was reached in 1962–64, when menhaden and mixed trawl-caught fish contributed about equally to the catch, with maximum landings of about 80,000 metric tons in 1962 (Figure 5-3). The subsequent decline, which began after 1964, was catastrophic, and the state no longer has a fish meal and oil processing industry.

Detailed landings, by species, show that the declines have been even more dramatic than the grouped data suggest.[2] In each of the three cases illustrated above the decline in landings would have been much more abrupt and total catches would have fallen to much lower levels if the industry had not

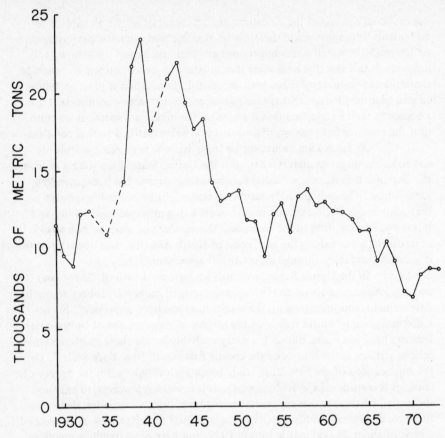

Figure 5-1. Landings of food finfishes in New York State 1929 to 1973 inclusive, by weight

constantly shifted to new resources as catches of traditional species fell. The reasons for some declines are known, the causes of others can only be guessed at, but the conclusion is clear that management of the marine fisheries of New York State has been ineffective. Fisheries for endemic resources like molluscan shellfish, which the state presumably could manage unilaterally if it had the incentive and the means, have fared scarcely any better than the others. A similar study of the marine fisheries of New Jersey, now underway, clearly points to the same conclusions.[3] Much the same could be said of any state along the Atlantic coast. The inability of the individual states to manage even their own endemic living marine resources shows that the real problem is a fundamental weakness in government, which cannot be brushed aside by placing the blame on a lack of cooperation between the states or on foreign

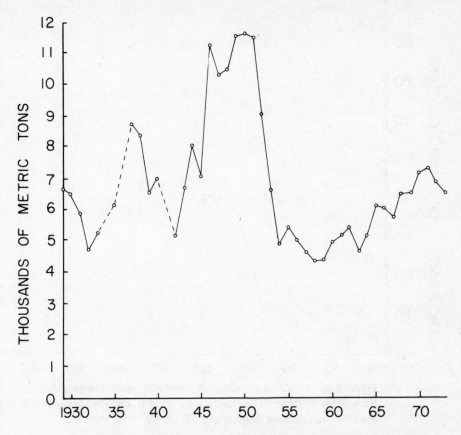

Figure 5-2. Landings of crustacean and molluscan shellfish in New York State 1929 to 1973 inclusive, by weight. Shells of mollusks not included.

fishing. These external issues merely aggravate the problem. Most other countries have done no better.

Recreational fisheries also are important in the coastal zone of the United States, as in some other countries. Data on saltwater sportfish catches are scanty. Nationwide estimates are available for 1960, 1965, and 1970 based on sample surveys by the Bureau of the Census. Usually it is believed that catches estimated from these surveys are overestimates, whereas official statistics on commercial landings probably are underestimates. Despite these uncertainties, however, there is little doubt that recreational catches are large, and that for some species the sport catch far exceeds the commercial catch. The saltwater sport fisheries are virtually unregulated in many coastal states of the

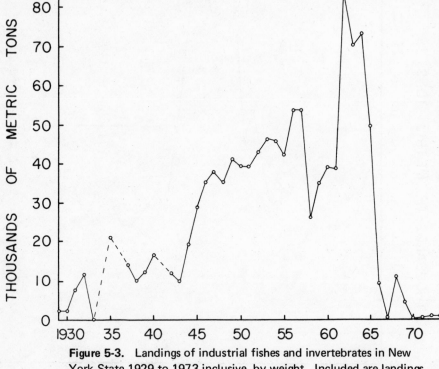

Figure 5-3. Landings of industrial fishes and invertebrates in New York State 1929 to 1973 inclusive, by weight. Included are landings used for manufacture of meal and oil, animal food, and bait.

United States, and many, New York included, do not even require that salt-water sport fishermen be licensed. Management of the coastal fisheries will require, as a minimum, adequate statistics on catch and fishing effort for recreational and commercial fisheries and management measures that apply with equal force to both segments of the domestic marine fisheries. This has not been achieved for the domestic marine commercial fisheries, and the prospects of success are not too bright. The possibility of extending adequate controls to marine sport fishing is even more remote. Some of the problems have been discussed in an earlier paper.[4]

Statistical areas for reporting domestic commercial and recreational catches and foreign catches in the North and Middle Atlantic coastal regions of the United States are not identical (Figure 5-4). Domestic commercial landings are reported by states, usually by point of landing rather than by waters in which the catch was made. Marine sportfish catches are reported by larger

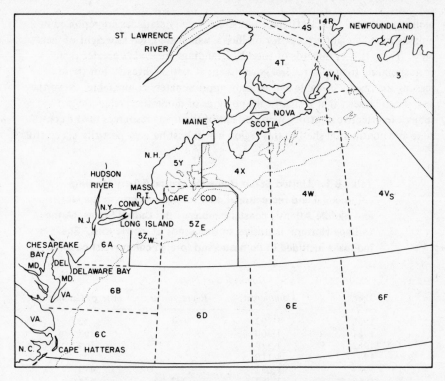

Figure 5-4. North Atlantic and Middle Atlantic regions of the United States Coast, showing statistical areas and place names mentioned in the text. Statistical subareas 3 to 6 off the coast are those adopted by ICNAF. Saltwater sportfish statistics are grouped into two regions: North Atlantic, which includes Maine to New York; and Middle Atlantic, which includes New Jersey to Cape Hatteras. These regions correspond approximately to ICNAF subareas 5 and 6 respectively. Domestic commercial landings are reported by states. The 200 meter isobath, or edge of the continental shelf, is indicated by a dotted line. Little if any fishing is done farther to sea.

regions, in which New York was lumped with the New England states as the North Atlantic region, and coastal areas from New Jersey to Cape Hatteras inclusive were grouped as the Middle Atlantic region. Foreign catches are reported in ICNAF Statistical Bulletins by subareas, of which subarea 5 corresponds approximately to the North Atlantic region of the saltwater sportfishery statistics, and subarea 6 approximates the Middle Atlantic region. The difference is that ICNAF includes New York coastal waters in subarea 6, which is essentially the same as the Middle Atlantic region of the marine angling surveys. ICNAF statistics are reported as live weight in metric tons, including shells of mollusks. United States official commercial fishery statistics are reported in pounds round weight, except for mollusks, which are given as weight of meats only. Thus, ICNAF records of total U.S. landings are always greater than corresponding figures in federal government statistical digests. For these reasons the figures in Table 5-1 are only approximately comparable. Nevertheless, it can be seen that if significant portions of domestic catches are not subject to effective controls, management of the living resources of the coastal zone and continental shelf in this region can at best be only partially successful.

Table 5-1. United States domestic marine fishery landings, commercial and recreational, and foreign catches, in the North and Middle Atlantic coastal regions of the United States, Maine to Cape Hatteras inclusive, in thousands of metric tons. Shells of mollusks included in domestic and foreign commercial catches.

Year	Domestic Landings		Foreign Catches
	Commercial	*Recreational*	
1960	1,292	164	32
1961	1,300		107
1962	1,379		264
1963	1,116		301
1964	1,032		413
1965	1,053	202	576
1966	933		572
1967	871		458
1968	875		625
1969	818		602
1970	985	233	429
1971	967		620

CAUSES OF DECLINING DOMESTIC FISHERIES

It is popular to blame foreign fishing for the problems of American coastal fisheries. It cannot be denied that foreign fishing over the continental shelves of North American is creating difficult problems by competing with American fishermen for traditional resources, sometimes interfering with domestic fishing operations, and contributing to serious overfishing of some stocks. These problems are particularly severe off Alaska, the Pacific Northwest, and New England. Bu ' the fundamental causes of the problems of domestic fishermen are social-political and economic, not biological. It does not require much research to find supporting evidence for this conclusion. Table 5-1 shows that the decline of the New York food finfish industry began about the beginning of the Second World War and accelerated when the war was over. The postwar

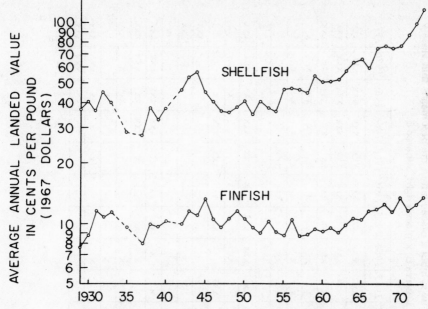

Figure 5-5. Average annual landed values of fishes and shellfishes in New York State 1929 to 1973, converted to 1967 dollars in terms of the wholesale price index for farm products computed by the Bureau of Labor Statistics, U.S. Department of Labor.

expansion of foreign fishing in the Nortnwest Atlantic did not begin until 1957, when the USSR began fishing in ICNAF subarea 3, off Newfoundland. Until 1960, only five nations were fishing in subarea 4 (Table 5-2), and stocks of fish south of Cape Cod were not threatened in any significant way until the Soviet fleet moved into subarea 5 in 1961 and into subarea 6 in 1963. As Table 5-2 illustrates, these moves were followed by southward shifts of fishing effort by several nations which had been fishing off North America for a long time, some for centuries, and by entry of new nations into the fishery. Until 1963, only one nation, the United States, had been fishing in the area now known as ICNAF subarea 6. In 1971 fleets of ten nations were operating in this subarea. By 1961 New York landings of food finfishes had already dropped to 55 percent of the 1939 maximum (Figure 5-1).

A clue to the causes of the decline may be found in the historical record of prices paid to New York fishermen for their catch of fish and shellfish used as human food (Figure 5-5). Average annual landed values per pound since 1929 suggest that there were at least five distinct periods in the foodfisheries of the state. Converted to standard dollars (1967 dollars), prices were relatively high at the time the great economic depression of the 1930s began, fell as the

Table 5-2. Nations fishing in ICNAF subareas 4, 5, and 6, 1953 to 1971. The symbol x placed left in the box means that vessels of that nation made catches in subarea 4 in that year; x in the center means subarea 5; x placed right, subarea 6.

	1953	1954	1955	1956	1957	1958	1959	1960	1961	1962	1963	1964	1965	1966	1967	1968	1969	1970	1971
U.S.A.	xxx	xxx	xxx	xxx	xxx	xxx	xxx	xxx	xxx	xxx	xxx	xxx	xxx	xxx	xxx	xxx	xxx	xxx	xxx
Canada	xx	xx	xx	xx	xx	xx	xx	xx	xx	xx	xxx	xxx	xxx	xxx	xxx	xxx	xxx	xxx	xxx
Spain	x	x	x	x	x	x	x	xx	x	x	x	xx	xx	xx	xx	xx	xx	xx	xxx
France	x	x	x	x	x	x	x	x	x	x	x	x	x	x	x	xx	xx	x	x
Portugal	x	x	x	x	x	x	x	x	x	x	x	x	x	x	x	x	x	x	x
U.K.								x		x									
Italy								x											
USSR									xx	xx	xxx	xxx	xxx	xxx	xxx	xxx	xxx	xxx	xxx
Norway									xx	x	xx	xx	xx	x x	xxx	xxx	x		
Non-members																			
Non-member A																xxx	xx	xxx	xxx
Non-member B																		xx	xx
Non-member C																			x
Poland										x	x	x	xx	xx	xx	xxx	xxx	xxx	xxx
Germany											x	x	x		x	xxx	xx	xxx	xxx
Denmark												x					x	x	x
Iceland												x				x	x		
Romania															x	x	x	xx	xxx
Japan																		xxx	xxx

depression deepened, rose in response to the demand for protein during the war, fell after the war at a time when costs were rising rapidly and some traditional resources were declining in abundance, and did not begin their present upward climb until the middle 1950s.

These trends were much more pronounced in prices of shellfish than food finfish. Average prices paid to New York fishermen for fish for human consumption have not changed very much for the past twenty five years. It appears that New Yorkers do not care very much where their fish comes from. Total landings of marine foodfish and shellfish in the state would provide less than 2 pounds per capita per year for the residents of the Greater New York Metropolitan Area alone. This area probably consumes considerably more than the 11-12 pounds per capita average for the nation as a whole. Thus, it is certain that most of the fish and shellfish that New Yorkers eat comes from other states and other nations. When a species becomes scarce locally, substitutes quickly move in to replace it. When the price is too high, a cheaper substitute apparently is used. It probably would not matter very much to the people of the state whether it had a commercial marine fishery or not, and this appears to be reflected in a lack of official interest by the state administration in management and development of its marine fisheries.[5] The situation in New Jersey and Delaware appears to be similar.

These conclusions are at present tentative, for the studies on which they are based are incomplete. They sustain the preliminary view that the problems of the fisheries of the coastal area of the United States known as the New York Bight are of relatively long standing, that the causes are economic and social-political, that the Second World War was an important factor, and that the solutions must come principally from domestic rather than international action.

ALTERNATIVE SOLUTIONS TO COASTAL FISHERY PROBLEMS

Contrary to the tentative conclusion reached above, public opinion in the United States, as already pointed out, favors extended jurisdiction over coastal fisheries. The United States position for the 1974 Law of the Sea Conference (LOS III), with respect to fisheries, which apparently is still being formulated, has evolved from strong adherence to the 3-mile concept, to grudging agreement on a contiguous zone to extend not more than 12 miles seaward from the baseline, to extension of fishery jurisdiction to 12 miles, and presently to management of fisheries by species or stocks rather than zones. Other nations will favor zones of fixed width, some pushing for broad jurisdiction, some narrow. The biological consequences of alternative regimes, whether the coastal state will control a broad or a narrow zone, will require equally clear statements of policy and assumption of responsibility at national and international levels, and equally effective data gathering and management measures.

Fisheries of the High Seas

I am not aware of any serious opposition to the concept of freedom of the high seas. The 1958 Convention on the High Seas provides, *inter alia*, for freedom of fishing in this area by coastal and noncoastal states. If this is a workable concept, then it might be argued that, since most nations agree that some parts of the ocean should not be included in the territorial waters of any state, thus being open to all nations, it follows that the area designated as high seas should include as much of the world ocean as possible. From the viewpoint of fishing this would appear to be a valid argument, for on the basis of present scientific knowledge and technology most fishery resources are over the continental shelves and in the coastal zones.

Agreement on a broad territorial sea or contiguous zone that includes most or all of the continental shelves would seriously restrict, if not preclude, access to most conventional fishery resources. Under such a regime freedom of fishing on the high seas would be meaningless. Since most, if not all, nations have been equally as unsuccessful as the United States in managing fisheries and fishery resources in waters entirely within national jurisdiction, proposals to broaden that jurisdiction should be viewed with suspicion by nations that wish to exercise their rights to fish on the high seas. Past performance by coastal states, as compared with the performance of international fishery bodies, suggests that unilateral jurisdiction would be less likely to manage the living resources wisely and protect the interests of other states in gaining access to those resources.

If this argument is valid, it follows that jurisdiction by the coastal state should be as narrow as possible, since control by international arrangements would in the long run appear to be the regime most likely to serve the interests of all states, including the coastal state.

Narrow Territorial Control

A logical extension of the argument in the previous section would lead to the conclusion that coastal states should have no unilateral jurisdiction at all over marine fisheries. However, the 3-mile territorial sea, and more recently an additional contiguous zone of at least 9 miles, have gained such precedence in international law that the probability of extending the high seas to include waters within 12 miles of the coast is about zero. Other compelling reasons, having nothing to do with fisheries, require that maritime nations have a zone of exclusive control off their seacoasts.

The management regime for living resources of the high seas need not be a monolithic multilateral United Nations–type organization. In fact, experience seems to have shown that the success of international fishery management has been roughly inversely proportional to the numbers of nations involved in the agreement, as Herrington, Clingan, and Wakefield have pointed out.[6] Instead, it would seem perfectly feasible, and probably much more successful, to establish regional or problem-oriented commissions with small membership.

If management is to succeed, however, much more importance must be attached to the obligation of the coastal state to establish equally effective management mechanisms within its own territorial waters. In most existing international fishery conventions the importance of this obligation has been underrated, or the commission established to implement the convention has ignored the matter. One strength of the Principle of Abstention is the mandate it places on the coastal state to manage living resources under its control, and the stringent criteria that must be satisfied if a stock or stocks of fish shall continue to qualify for abstention by one or more member states. Incorporation of the Principle of Abstention in the Convention on the High Seas Fisheries of the North Pacific Ocean forced the United States and Canada to take extraordinary (and costly) measures to manage the major salmon stocks and certain other living resources within their territorial waters. For the United States, at least, these are the only major fishery resources that are under reasonably effective management regimes within territorial waters.

200-Mile Zone

A 200-mile territorial sea, or zone of fishery jurisdiction, has some biological basis. This is roughly the average width of continental shelves around the world, and most living resources presently being harvested live on or over the shelf. However, 200 miles is not an ideal regime for fishery resources because the shelf varies greatly in width and because some resources migrate well beyond 200 miles.

A 200-mile zone of jurisdiction over fisheries can be established by unilateral action, as some countries have done, or by international agreement. Unilateral action is not acceptable from a biological point of view for the reasons already discussed. At best it carries with it only a moral obligation to manage the living resources and the fisheries that harvest them, and there is little if any evidence that nations acting alone have the incentive or the means to manage their coastal fisheries even in much narrower zones of jurisdiction. International agreement on 200 miles or some other regime based on broader jurisdiction by the coastal state, since it presumably would include mandatory provisions for management, would appear to be superior to unilateral action.

Management by Species or Stocks

A proposal that originated in the United States, in which the coastal state would have jurisdiction over species or stocks of fish off its coasts wherever they might go in the ocean, is at first glance attractive because it is logical scientifically. Such a regime is not likely to receive wide support because it is in effect a slightly modified version of the Principle of Abstention. For all its merits, the Principle of Abstention is generally unacceptable because it is a form of exclusion, which was rejected in 1958 and is much less likely to be adopted in 1974.

Management by species or stocks is also inconsistent with developing thought in scientific fishery management. Managers of fisheries in small lakes and ponds have long known that selective fisheries, regulated species by species, usually do not respond in the way that simple management models predict. The well-known reversal in abundance of Pacific sardine (*Sardinops sagax*) and anchovy (*Engraulis mordax*) through interaction of selective fishing and environmentally induced changes in abundance, and the complicated changes in fish stocks in the Great Lakes from selective fishing, environmental changes, and introduction of exotic species, have made it clear that the ocean is not so large and its living resources so vast that man cannot alter the biological equilibrium there also. Most fishery scientists now believe that the ultimate goal must be management by biological communities. This takes into account the transfer of energy from one species to others as the effects of fishing and environmental variables interact. This raises the theory of fishing to a much more complicated level than the theory based on management stock by stock, and may vastly increase the difficulty and cost of wise management.

No Formal Regime

One alternative that usually is ignored in discussions of fishery management is to consider the consequences of no regime at all. In the United States, although we place great importance on conservation and management, our performance in marine fishery management in territorial waters has been far from successful. Despite strong public opinion to the contrary, we have applied the principles of scientific management much more successfully when we have entered into formal agreements with other nations. This probably is generally true for most major fishing nations.

Most successful distant water fishing nations have operated in a way quite contrary to the conventional principles of scientific fishery management. In a sense they have been "mining" the living resources, identifying areas of high biological productivity, fishing those resources until catch rates became unprofitable, constantly exploring new areas and new stocks to open up new fishing grounds. In this the distant water fisheries have been aided by efficient intelligence systems, great flexibility, and capability to utilize a wide variety of resources for human food and industrial purposes. This approach is inconsistent with conventional methods of fishery management because it can lead to serious direct overfishing of some stocks, like haddock (*Melanogrammus aeglefinus*) on Georges Bank, or indirect overfishing by taking incidental catches of species not deliberately sought, like halibut (*Hippoglossus stenolepis*) in the Northeast Pacific. On the other hand, from the standpoint of continued large catches the distant water fleets have a great advantage over land-based local fisheries, with the additional advantage that scientific management for maximum sustainable yield (or any other concept of maximum returns from fishing) is not a very important matter to distant water fishing nations until the total world harvest approaches the maximum sustainable yield of total usable biomass. In the United States it is popularly believed that this is a wasteful and unacceptable method of fishing, which is true from the standpoint of the coastal state whose traditional coastal fishery resources are affected. From the broader perspective of world fisheries it is possible that the strategy of distant water fishing is basically sound. This does not help to resolve the controversy between distant water and coastal fishing nations, and this is the basis of the argument that coastal states should have some form of preference or priority over the living resources of their coastal waters.

In the United States the cost of fishery research, development, and management, compared to the return to the economy from fishing, is relatively high. No one has made a detailed study of expenditures by federal and state governments on fisheries, but it is probably conservative to estimate that the federal government invests at least $60 million a year and the states together another $60 million. The landed value of the catch is slightly over $700 million at current prices, and from this must be subtracted the costs of catching, probably an indeterminate figure, but certainly relatively large. Thus, the subsidy to the fishing industry in the United States is substantial, and in the view of some economists it brings no net return to the national economy. The net effect of these expenditures probably has been nothing more than to slow the rate of decline of traditional domestic fishery resources. The benefits to the resources, if not to the fishermen, as already recognized, have been better in the international arena. Some restrictions on coastal fishing are necessary, such as preserving adequate spawning conditions and allowing access by anadromous species, and reasonable limits on catches of highly vulnerable resources like stocks of shallow-water mollusks. It would be interesting to examine the consequences of eliminating governmental expenditures on domestic fishery management, at least for migratory marine species, and either allocating the funds as a direct subsidy to fishermen, or allowing international fishery bodies some voice in their administration.

Limitations on Entry

Limited entry has been advocated as the only economically sound way to limit the catch and provide a return on investment to the fisherman. The concept should be attractive to fishermen, for it should reduce or eliminate economic waste. The economies of major fishing nations are sufficiently different, however, that it would be very difficult to arrive at an agreed-on formula. As a practical matter, limited entry as it is conceived in North America has no place in international arrangements. If agreement can be reached on the maximum allowable catch and national shares of that catch, economic refinements such as limited entry can be handled more effectively under domestic arrangements, each nation choosing how much capital and labor it wishes to invest in fishing. The Japanese system of licensing its high seas fishing operations is a form of limited entry.

The explicit goal of maximum sustainable biological yield, as incorporated into most existing international fishery agreements, has been challenged by some economists. From a purely economic point of view the argument is valid. However, several other circumstances have a bearing on the matter. One, the difference in national economic structures, has been mentioned above. Another is that if the end objective is to take the maximum sustainable yield of protein from the sea, a management regime based on maximum economic yield, although it certainly should bring a larger return in dollars to the fisherman, probably will require a lesser harvest of protein than the maximum possible. Even more important to Western economies is the often overlooked fact that maximum economic return to the fisherman may not be maximum economic return to the processor or distributor of fishery products, to whom the marginal yield represented by the difference between the catch for maximum economic yield to the fisherman and the maximum biological yield may represent the middlemen's margins of profit. Limited entry, since for most species it would require a lower rate of removal than that which would produce the maximum sustainable biological yield, could provide a useful safety factor against overharvesting. Desirable though it may be by some criteria, limited entry probably will not be a factor in international fishery agreements.

SUMMARY AND CONCLUSIONS

Most of the previous discussion applies generally to world fisheries. This conference is concerned with a more restricted part of the world ocean, the North Atlantic, which is a special case because most of the major fisheries are already under international scrutiny and management under the provisions of conventions which span the North Atlantic from North America to Europe.

The biological consequences of the various regimes that probably will be discussed in LOS III, and some regimes that probably will not receive consideration, have been viewed in the light of past performance, domestic and international, with emphasis on the region known as New York Bight. Domestic performance in management of coastal fisheries under national jurisdiction has not been impressive. The states concerned have not even succeeded in maintaining equilibrium yields of endemic resources, unavailable to foreign fishermen, by unilateral or joint action. International management, through ICNAF, has been fraught with many difficulties, and has not succeeded as well as most parties would have desired. Yet slowly the commission has come to grips with some of the most difficult problems, and although much remains to be done, recent progress has been encouraging.

The same general conclusions probably can be drawn with respect to fishery management by the United States within the 12-mile zone of fishery jurisdiction in almost all parts of the seacoast as compared with the performance of international fishery commissions of which the United States is a member. It is probable that much the same state of affairs exists among all the major fishing nations, and this is why the United States has not been singled out for criticism for its contrary policies and performance at home and on the high seas.

Attractive though the prospect of an extension of jurisdiction by coastal states over fisheries and fishery resources on the high seas may be, the record of performance is not reassuring. Certainly, the possibility of unilateral extension of sovereignty or proprietorship does not appear to be a viable solution. Extension of jurisdiction by international agreement is far more appealing, because it is presumed that such agreement would include mandatory provisions for research and management, which would obviate reliance on the historically less successful method of moral obligation. Although it has no biological implications provided that adequate management mechanisms are available, the concept of preference or priority for the coastal state has some merit for other reasons. However, the argument for exclusive jurisdiction, while it has attractive short-term features, may not be the best strategy in the long run.

In the North Atlantic Ocean potentially workable international arrangements are already in existence, particularly in the form of the International Commission for the Northwest Atlantic Fisheries (ICNAF) and the North-East Atlantic Fisheries Commission (NEAFC), but with the aid of the International Council for the Exploration of the Sea (ICES), and the International Commission for the Conservation of Atlantic Tunas (ICCAT) and the International Whaling Commission (IWC) for certain high seas fisheries. These arrangements may not be ideal, but they have been far from ineffective, and they should not be discarded lightly. With some reconstitution, the details of which must be worked out by those most familiar with the workings of these commissions, they can continue to improve their performance and provide scientific leadership for management. Under the circumstances, extension of national jurisdiction would appear to offer little promise of improvement in present arrangements. Past experience suggests that the consequences of extending national jurisdiction in this area might be to add new complications and perhaps reduce the chances of success.

Experience also seems to show that the other alternative, a single monolithic international body, would be undesirable. The principle that fishery problems are best solved by those directly concerned seems to be borne out by the success of the bilateral fishery commissions as compared with those that have large numbers of members. But the increasingly complex concepts of scientific fishery management, and the growing interactions between fisheries, have made the simple concepts of biological management of a few years ago obsolete, and more complicated arrangements are called for today. Probably the best compromise will be broad general agreement on world fishery management policy, implemented by a complex of regional commissions, subdivided when appropriate, as ICNAF is, by smaller ocean regions, with provision for special handling of unusual or unique problems.

These observations appear to emphasize the extreme importance of international agreement at LOS III on a fishery management regime that will be acceptable to most nations. Failure to reach international accord almost certainly will bring about an acceleration of extravagant unilateral claims which are highly unlikely to be favorable to the health of the living resources. For the North Atlantic Ocean the arrangements may not need to be very much different from those in existence already.

Notes to Chapter 5

1. J. L. McHugh, *Marine Fisheries of New York State*. U.S. Dept. Commerce, Natl. Marine Fish. Serv., Fish. Bull. 70, 3 (1972): 585-610.

2. *Ibid.*

3. J. L. McHugh, "Jeffersonian Democracy and the Fisheries," *World Fisheries Policy, Multidisciplinary Views*, ed. Brian J. Rothschild (Seattle: Univ. of Washington Press, 1972), pp. 134-55.

4. Jay J. C. Ginter, "Marine Fisheries Conservation in New York State: Policy and Practice of Marine Fisheries Management" (A report to the New York State Assembly Scientific Staff and the New York Sea Grant Program, Marine Sciences Research Center, State Univ. of New York, Stony Brook, N.Y., 1974).

5. W. C. Herrington, Thomas Clingan, and Lowell Wakefield, "Marine Life," *Public Policy Toward Environment 1973: A Review and Appraisal, Ann. N.Y. Acad. Sci.* 216 (1973): 95-104.

Chapter Six

A Market for Fishery Resources?

Francis T. Christy, Jr.

The views of the North Atlantic countries toward the living resources of the sea provide an intriguing anomaly. These countries, in their ordinary affairs, are strongly oriented to the fundamental rights of property and to the use of market mechanisms as a basis for the allocation of resources. But when it comes to marine fisheries, the concept of property is accepted reluctantly, if at all, and market mechanisms are generally rejected. Unlike the countries of the "third world," the North Atlantic states continue to worship at the altar of Hugo Grotius and continue to chant the litany of the freedom of the seas, unwilling to face the fact that there is no such thing as a free seafood lunch.

It is not difficult to understand why property rights have been so assiduously avoided in the past. The freedom of the seas is a cherished principle which has been of considerable value to those maritime states that have been able to make the most use of it. Removing the principle, and establishing property rights, has been difficult—not only because swimming fish do not respect man's borders, but also because the action necessitates a direct confrontation over the distribution of the seas' wealth. As long as other opportunities for fishing were available and no one's ox was unduly gored, the states preferred to compete under the condition of free and open access rather than undertake conflict-inducing negotiations over the explicit distribution of wealth.

But these conditions have changed dramatically. Unilateral extensions of jurisdiction by numerous southern hemisphere states have reduced the opportunities for taking fish in other waters. Major increases in size of fishing ventures and improvements in the technology of fishing have led to the goring of many oxen. And the costs of maintaining the freedom of the seas have greatly increased, as both biological and economic waste have become more pervasive and more severe. The North Atlantic countries, whether they like it or not, can no longer avoid facing up to the difficulties of dismantling the principle of free and open access and of acquiring some form of property right over the resources.

Iceland has done so by unilaterally extending its jurisdiction to fifty miles. In the Northwest Atlantic, the property rights—such as they are— have been acquired by the distribution of national quotas. In both cases, a first step has been taken, albeit reluctantly and at the cost of certain units of good will. But there appears to be little move to take the additional steps necessary for the establishment of effective markets for fishery resources. Property rights have been created, but without provisions for transferability. Rights to fish are determined by the negotiating skills of diplomats rather than the workings of an economic market.

Iceland could, if it wished, open a market for its resources rather than indulge in a primitive barter of numbers of vessels with a right to fish. It could follow the model of many Latin American states and permit foreign fishing subject to the payment of license fees or royalties. Or it could follow its own model in the rental of rights for recreational salmon fishing in its rivers. The British, now that it appears that they have accepted Iceland's jurisdictional claim, would not necessarily find such license fees more onerous than reductions in the number of vessels or in allowable catch. And Iceland would be receiving revenues not now available.

In the Northwest Atlantic, the national quotas that have been agreed upon provide a form of property right, but one that is severely limited by the decision that the quotas are not transferable. This precludes a market for the resources. It would not be difficult to reverse the decision. Indeed, the original quota agreement on herring specifically permitted transferability. If this were readopted and made available for all stocks, states could transfer their quotas, or shares of their quotas, by sale, lease, or trade. Those states with quotas too small to be used effectively could receive some benefit by leasing them to others or could acquire shares from others sufficient to warrant a fishing venture. The market would permit a flexible response to changes in demand and supply functions. It would also facilitate entry by new states. Under the present system, new states can participate only by successfully claiming some of the portions reserved for newcomers or by getting member states to reduce their quotas. With a market, they would be able to purchase their way into the fishery. A market would also improve the allocations of capital and labor, since the monetary costs of acquiring the resources would be internalized in the fisheries. In short, transferability of quotas is an important step in improving flexibility and efficiency.

Opposition to the creation of markets for fishery resources appears to be very strong, however. Some of the reasons for the opposition are clear, but there may be several other, and perhaps more important, reasons that are not well understood at present. Further understanding is eminently desirable because the presence of newly acquired property rights provides, for the first time, a real opportunity to establish markets and rationalize fishery industries. Whether or not such steps should be taken depends upon the costs and difficulties that are likely to be incurred in the process. And these costs can be evaluated only if the nature of the difficulties is carefully identified and made explicit.

Some possible reasons for the opposition to markets for fishery resources are discussed below. Several of these, however, relate to cultural patterns and behavior and should properly be analyzed by cultural anthropologists and social psychologists. Thus, the most that can be done here is to entertain some speculations about the reasons, and to suggest that more research by competent anthropologists and others is extremely important.

One of the reasons for opposition to markets lies in the opposition to the establishment of property rights. For example, British opposition to Iceland's claims might be more deeply undermined by an acceptance of license fees than by a limit on the number of vessels. In the Northwest Atlantic, the major, and quite dramatic, efforts taken to negotiate the national quotas apparently left little time or opportunity to evaluate the consequences of the additional step of transferability (though having done it for herring, it is a bit surprising that transferability was apparently rejected for the other stocks). But both these reasons are transitory. The British are now moving much closer to the adoption of extended limits of jurisdiction for themselves, and the states involved in the Northwest Atlantic can now devote some study to the possible creation of markets.

A more durable source of opposition lies in the fear that a market for fishery resources would be distorted by the differences in national subsidies and grants given to the fishermen. The centrally planned economies and the countries with heavy subsidies might outbid the others and acquire rights to an excessively large share of the resources. To the extent that countries view fisheries as a source of economic revenue, such developments are not necessarily damaging, since they might stand to gain more by leasing their rights than by engaging in fishing. If, however, fisheries are viewed primarily as a source of employment opportunities or as a source of protein (with little regard to cost), then it might require increasing subsidies to match those of the high-bid countries. This problem could be handled by placing constraints on the market. For example, Iceland might lease out only that tonnage of fish that it cannot catch with full employment of its fishermen.

Other possible reasons for opposition are more difficult to evaluate, although they may be of considerable importance. Nationalism is undoubtedly a strong factor. There may be as much opposition to foreign fishing in "our" waters as there would be to foreign tilling of our soil or exploitation of our minerals. Nationalistic attitudes in North Atlantic countries are likely to be less satisfied by the extraction of taxes or fees from the foreigners than by total (eventual or immediate) expulsion of the foreigners.

Opposition to a market by North Atlantic countries may also be due to the strongly ingrained principle of the freedom of the seas. Under this principle, the rule of capture prevails and only those who actually fish have a right to share in fisheries wealth. There is little or no experience with the concept of landlords in North Atlantic fisheries, and thus little experience with systems that permit the extraction of benefits by nonusers. The establishment of a market for resources that have hitherto been considered free represents a significant departure from tradition, even though markets are taken as given for almost all other natural resources. There is likely to be considerable un-willingness to consider a market mechanism as a viable alternative to techniques that are more in keeping with tradition.

An additional reason which might be mentioned is the fear that the establishment of a market would be detrimental to the personal interests and values of individual fishermen. The availability of marketable property rights would make fisheries more attractive to "big business" than they are at present. This might threaten the independent way of life of many fishermen and might subject them to conditions which they may well find discomforting.

It is admittedly difficult to evaluate the relative importance of these, and other, reasons for the opposition to the establishment of a market for fishery resources. Nevertheless, the effort to do so would be extremely worthwhile and timely. For the first time in the North Atlantic, the states are acquiring a degree of property rights over the resources. And there is now an opportunity to move further and to take steps that would facilitate the eventual creation of a satisfactory market. The next steps that can be taken are small, and would not require massive administrative machinery. Iceland, having limited the number of foreign vessels, could easily impose license fees or taxes on catch. These would not necessarily be burdensome to the foreigners, since the present limitations on fishing effort may lead to surplus economic returns. The states in the Northwest Atlantic fisheries could take the easy step of reintro-ducing the right to transfer national quotas, or shares of the quotas. While there might be some costs in keeping track of the transfers, these would be more than offset by the reduction in need for prolonged negotiations.

The major impediments to taking such steps appear to derive from cultural and social traditions and patterns of behavior. It may be that these traditions and patterns of behavior are important enough to society to warrant a decision restraining the creation of a fishery resource market. But at the moment, we do not know.

Chapter Seven

The Case for a North Atlantic Preferential Fishing System

Austen Laing

THE FORMS PREFERENCE MAY TAKE: OPEN AND DISGUISED

A preferential fishing system is commonly understood to mean one which accords preferences to the coastal state or states in a regulated high seas fishery, i.e., in an international fishery that lies wholly or partly outside national fisheries jurisdiction or, what is for all practical purposes the same thing, are what would lie outside of exclusive economic zones supposing such zones became recognized under international law.

    There are many ways in which such preferences may be expressed: partial or total exemption from restrictions applied to other participating states in respect of gear, type of fishing or vessel, or of area, or seasonal closures, or of bycatch limitations; or, especially if it is important to present the system in conservation terms, the restrictions may apply to all participating states equally in form but unequally in fact and in such a way as to favor the coastal state. For example, the prohibition recently adopted by ICNAF on fishing with bottom trawls by vessels over 145 feet in a triangular area off Cape Cod is nondiscriminatory in form, but in practice it allows U.S. coastal vessels to go on fishing for demersal species while barring others from that area.

    More usually, however, the preference is expressed as a proportion of the total allowable catch (TAC) reserved for the coastal state(s), in addition to the share to which it is entitled as a result of the application of common criteria, or some agreed formula, to all participants in a catch limitation scheme (CLS). Such a preference could, of course, be expressed in terms of effort if measures for the management scheme included effort limitation.

Even with common criteria or an agreed formula, there can be disguised preferences to the coastal state which add to the overt preference accorded to it. These arise from the choice of criteria or the elements of the formula for the apportionment of the fishery cake. A not unusual proposal is for the portion of the TAC not specially reserved for the coastal state to be divided in the ratio of the catches of the participants over some agreed period with some system of weighting of its constituent parts. But the choice of both the period and the weights may significantly affect the share-out of the unreserved portion. (Even greater opportunities for disguised preferences exist where effort limitation is involved.)

A STANDARDIZED FORMULA FOR PREFERENCE

Attempts have been made by some members of both NEAFC and ICNAF to standardize the formula to be used for the total share-out. The 40/40/10/10 formula has gained the greatest currency; this gives a coastal preference of 10 percent of the TAC, makes a similar reservation for newcomers and growth with half of the remainder divided among all participants in the ratio of their catches during the preceding ten years and the other half in that during the preceding three years.

All such attempts have largely if not wholly failed. It is fair to say that the formula quoted has provided some sort of precedent, but only to provide the starting point of international negotiation: final share-outs have not usually borne any close likeness to that which the application of such a formula would have produced. But then it may be questioned whether a uniform preference of 10 percent of the TAC could possibly be fair when the circumstances of coastal states, as well as of others, can vary so widely. Leaving aside all other considerations, a 10 percent preference means one thing when the coastal state is and has been taking only 15 percent of the total catch and it is something quite different when the proportion is 75 percent. (But, unfortunately, not everyone agrees about the significance of the difference!) It may also be asked whether an apportionment formula is right to give (a) added weight to recent years when it is often the upsurge in these years that has necessitated a CLS and (b) a double share to those who have caught twice as much as others when they can be held to have caused twice as much damage to the stock!

PREFERENCE VERSUS PRIORITY

Almost from the start, however, the formularistic approach has been in conflict with the view that the coastal state should be allowed to take as much of the TAC as it is capable of catching. The conflict has not been formally resolved though, implicitly, the latter approach has been conceded on an increasing scale during the last two years or so but particularly in 1973. Provided the catch claimed by the coastal state is felt by others to be "reasonable" in relation to its recent catches (and what is considered "reasonable" tends to grow with the passage of time), the application of the formularistic method has not been pressed: the main arguments have taken place over the size of the TAC, with the needs of conservation made somewhat, but not wholly, subservient to the desires of all concerned to obtain—chiefly in absolute terms—an acceptable catch entitlement.

DE FACTO ACCEPTANCE OF PRIORITY: ICNAF, NORTHEAST ARCTIC AND FAROE

This *de facto* acceptance of coastal state priority, as it may be termed to distinguish it from other forms of coastal state preference, was illustrated at the 1973 meetings of ICNAF when the Canadian and U.S. claims for taking as much as they could catch (in specific rather than general cases) were hardly resisted: in 1974, for the area off the New England coast of the United States, an overall TAC of 924,000 tons was agreed upon—as against 800,000 recommended by scientists. The U.S. share is about 2 percent higher than its estimated 1973 catch and about 2½ percent lower than its average for 1971-73, while Canada's is about 42 percent higher than its catches in either 1972 or 1973 and nearly 3 percent higher than its average for 1973-73. As the overall TAC is about 22 percent below the estimated overall catch for 1973 and the overall average for 1971-73, it is clear that the remaining countries have had to accept substantial reductions for 1974.

     In a disguised form, this implicit acceptance of coastal state priority is to be found in the agreement related to Northeast Arctic cod that is operative in 1974. The signatories are Norway, the U.S.S.R. and the United Kingdom; and, in deference to the administrative difficulties in controlling many thousands of small boats operating from numerous small ports in the remote regions of northern Norway, it has been agreed that Norway is not obliged to stop fishing, though it is hoped she would do so, when her agreed share of the TAC has been reached. If, however, Norway avails herself of this escape clause, then the

U.S.S.R. and the United Kingdom may free themselves of their obligations under the agreement. But, supposing Norway did exceed her share, the two other signatories would need to consider very carefully the full implications of following suit.

The same agreement contains a further point worthy of mention. The whole of the coastal preference is accorded to Norway despite the fact that the U.S.S.R. is also a coastal state in respect of the fishery concerned and may even have a coastline of comparable length to that of Norway. Admittedly, the proportions of the stock to be found in waters within median lines may make the U.S.S.R. the minor interest. But it has nowhere been suggested that the major coastal interest should take all. It is true that the stock spawns largely if not wholly within Norwegian waters, but it has yet to be accepted that for species such as cod—unlike migratory salmon and trout—the preference should be confined to the state in whose waters it spawns.

The losing battle being fought by formularistic preferences against coastal state priority was also exemplified in recent negotiations concerning the regulation of the stocks in the Faroe region. The resultant agreement, by requiring a cutback in catches of everyone but Faroe itself, can be said implicitly to have conceded coastal priority. The preferences—not only in the share of the TAC allotted to Faroe, but also in the restrictions on the size of vessels to be used and the timing of area closures as well as in the exemptions for Faroe vessels—were together great enough to be regarded as tantamount to priority.

PREFERENCE, PRIORITY, AND EXCLUSIVE JURISDICTION—ONE CONTINUUM

While there is an undoubted conflict between preference and priority, it might be misleading not to point out that—in practice if not in principle also—the two differ in degree rather than in kind. It has already been noted that the best known of the preference formulae earmarks 10 percent of the TAC for "new-comers and growth"; there can hardly be any doubt that, if negotiations ever got into such a vein, priority would be accorded to coastal state growth over all other claims for any goodly part of this 10 percent. Moreover, the concept of preference does not by any means preclude either an earmarked share for growth larger than 10 percent or a preference to the coastal state in the division of any such earmarked share. Indeed, it is not in my view inconsistent with the notion of preference, or of a formularistic expression of it, that such preference should steadily grow from one year to the next.

It has also been noted that recent practice has been to go a long way to meet the wishes of the coastal state to catch as much as it is able. Fundamentally, therefore, the practical conflict has been a question of pace: preference moves more slowly than priority, but, given continued coastal state growth, both eventually mean exclusive exploitation by the coastal state. Again in practice, if not in theory also, exclusive exploitation must give rise to exclusive jurisdiction. Hence, preference, priority, and exclusive jurisdiction belong to a single continuum or almost an inexorable line of development.

THE 200-MILE WHIP

Those who support the maintenance of exclusive fishing limits of narrow breadth would generally prefer a system of preference to one of priority and would generally elect for the latter as an alternative to extensions of jurisdiction. The first has given way to the second largely because in practically every specific case in the Northeast Atlantic in recent years the issue has been presented to the distant water states, sometimes but not always explicitly, as a matter of either yielding to the "reasonable" wishes of the coastal state(s) or facing early and substantial extensions of limits with implications going well beyond the coastal states concerned. The coastal states were not necessarily wrong to act in this way: the manner in which the International Fishery Commissions had habitually gone about their business, and particularly their being effectively tied to a unanimity rule, had until recently at least frustrated the development of truly efficacious stock management schemes. But, whatever the reasons or justification for it, the underlying threat of substantially extended limits—usually to 200 miles—has eroded the concept of preference to the advantage of priority, which in turn buttresses extended jurisdiction itself.

THE NARROW LIMIT ARGUMENT

It is not that the proponents of narrow limits, who may for present purposes be identified with the supporters of preference, entirely lack economic argument to support their views. It is at least arguable that the greater the opportunities for distant water fishing (or, more generally, for fishing by "stranger[1]" vessels), the better the possibilities for a beneficial international division of labor and for a high rate of both technological advance and product innovation: that is to say, the wider the access permitted to an area the greater the likelihood of speedy discovery of (a) hitherto undiscovered resources, (b) new uses and higher values for underutilized resources, and (c) more economical means of exploitation and of marketing. ("Wider" access should not, of course, be confused with freer access: whatever the regime, entry to a fishery needs to be controlled; the question here is simply one of who is to be allowed in.) In short, it can be argued that distant water fleets are important vehicles of fishery pioneering.

Furthermore, permitting regulated access to more than purely local fishing vessels—and, above all, to distant water fleets—means that the mobility of stranger vessels can be used to greater advantage, in adjusting the distribution of fishing effort to the constantly changing pattern of stock levels, than would be the case were entry to be totally denied to them.

It is almost inevitable, if not also right, that any reductions in fishing effort that may be required in a fishery or area from time to time should fall with greater severity, if not exclusively, on those vessels possessing greater mobility, i.e., the stranger and, more especially, the distant water fleets. To treat stranger vessels as the principal balancers in a fluctuating fishery has a considerable advantage for local fishermen: it reduces or even eliminates the need for them to vary their own effort in sympathy with changes in stock levels, i.e., they can remain fully or more fully employed, with their annual catches fluctuating less, than would otherwise be possible, while at the same time observing the precepts of good husbandry.

It may be countered that this advantage could be retained, while confining access to local fishermen exclusively, simply by underexploiting the fishery; this would give local vessels the added advantage of higher catch rates per vessel than those which would prevail were strangers admitted to the point where the fishery were fully exploited. But this runs into the objection that underexploitation is to be condemned on humanitarian grounds, because it unnecessarily restricts food supplies, if not on economic grounds also (depending, however, on the criteria of optimum exploitation). It is also to be questioned whether indigenous pressures towards expansion could be resisted as long as the maximum yield were not being taken.

It may also be countered that extended jurisdiction does not *per se* imply denial of access to foreign fleets. It is open to any coastal state to admit what vessels it pleases, admittedly on its own terms; but, in a world of greatly extended limits, uncertainty about continued access to waters under foreign jurisdiction must inevitably lead to a significant decline in distant water fleets and, hence, to a diminution in their "balancing" and "pioneering" roles, even supposing that coastal states generally were—uncharacteristically—not tempted to behave chauvinistically.

Mention should also be made of the economic dislocation and social distress which may be caused in those parts of distant water states heavily dependent on distant water fishing. This is not likely to be confined to distant water fishermen themselves but would probably extend to the ancillary trades and service industries, docks and harbors, processing, transport, and distribution generally, all of which would need to adapt to a changed pattern and levels of demand, on the one hand, and changed patterns and continuity of both supply and prices, on the other. It may be rightly objected that discontinuities of this kind can just as easily stem from the introduction and development of a preferential system as from that of coastal state priority or limit line extensions and that, in any event, the differences between them are those of degree, not of kind. It cannot be denied that phase-out arrangements can be a feature of all three, but the likelihood is that the length of the all-important period of adjustment will diminish as we pass from the first to the third.

The only serious qualification that needs to be made to this conclusion is largely confined to those places where the distant water state is not in the position of a suppliant pleading favors for old time's sake, but is itself a coastal state in the waters off whose coasts the roles are reversed. In such cases, and all fishing nations in Northwest Europe are fishing in each others' backyards, the negotiation of bilateral arrangements becomes possible—though it may rarely if ever be the case that the two parties are negotiating from positions of equality.

CONSERVATION

In the North Atlantic at least, the proponents of extended limits have usually advanced their claims in terms of the needs of conservation: the international commissions have shown themselves to be ineffective, therefore (*sic.*) stock management must be put into the hands of the coastal state. This ignores the conclusion reached by FAO[2] that "there appears to be little relation between the success or otherwise of management actions and the type of jurisdiction within which the resource lies. There are...at least as many examples of depleted resources which were under the control of a single country...as of those occurring outside national jurisdiction."

It may be said that national jurisdiction has not had a proper opportunity to demonstrate its worth because of the comparatively narrow limits which have generally been in force; a meaningful chance will not present itself until limits are greatly extended. This might be true, but it should not be forgotten that there have been opportunities even within narrow limits which have been conspicuously spurned.

Moreover, it needs to be added that when the issue is purely national, pressure groups get to work in directions which all too often are in conflict with the long-term requirements of the fishery. The government of the country concerned does not then have the discipline which comes from being obliged to counter the views of distant water states (behind which it may conceal its opposition to the views of its own fishermen). Having to face the latter openly it may falter, and when political expediency walks in through the door the needs of conservation fly out of the window.

Regrettably, a similar position sometimes obtains with fishery scientists who, being almost always government employees, tend to produce answers the government wants or, at least, omit to furnish relevant information on the grounds that it has not been requested. In their defense it may be said that they are bound to promote the national interest as defined by their political masters, whatever their scientific consciences may prompt. Moreover, they may lack the discipline imposed by having to work with scientists of other countries when the fishery is international and, in some cases, they may lack also the expertise which their overseas colleagues could provide.

Although the track records of **ICNAF** and **NEAFC** may be unimpressive, their recent sprints under the lash of a threatened 200-mile limit have demonstrated that international commissions are not of necessity as ineffective as they are often made out to be. (In any event, though some members of international commissions are not slow to deride the latter, it must not be forgotten that the commissions do not possess an existence beyond their membership and that their faults and shortcomings are, therefore, those of their members themselves.) That is to say, if it be said that national management has not had a proper chance to show its worth, it may be claimed with at least equal force that international management has not had a proper opportunity either: given the right legal framework and constitution it may well show itself highly effective.

This is an issue which needs studying because, in the Northeast Atlantic at least, a great need will remain for international cooperation whatever the breadth of fishery limits. A 200-mile limit in this region is almost synonymous with median lines, except to the west of Portugal and parts of Spain and France as well as of Ireland and Scotland. Many stocks would be common to the waters of more than one country and, moreover, there are problems of relationships between species which cannot best be solved on a purely national basis. Furthermore, the existence of a community demanding nondiscrimination between EEC vessels, while permitting fishery regimes within national waters to be determined by the member state concerned, makes for further complexity. Finally, almost every fishing nation of Northwest Europe has within its median lines stranger vessels of almost every other fishing nation of Northwest Europe. Some of the issues could be determined satisfactorily on a bilateral basis but many could not, and the need for an international commission, though with a new constitution and new powers, will remain.

PHASE-OUT

So much has already been conceded in the North Atlantic to the concept of coastal state priority that it would be flying in the face of reality to argue for a preferential system which did not incorporate an arrangement for the phase-out of all foreign fleets. The counterpart of this is, however, that it would be inequitable *not* to include a reasonable phase-out arrangement in any new fishery order for the North Atlantic. But it seems to me to strain the meaning of words to go on talking about a preferential system when it contains within itself the means of its transformation into an exclusive fishery zone. And, as the latter can hardly be separated from exclusive jurisdiction, it would be wrong to mince worlds: a case can be made out for a preferential system, but it won't stand up—events have overtaken it.

The reality is toward substantially extended jurisdiction and the questions are not so much as to breadth but, rather (a) the steps to be taken to ensure that foreign fleets are treated equitably and (b) the constitution and powers of the international commissions that will be needed to deal with the interrelationships between stocks which extend beyond limits however defined as well as such high seas fisheries as may remain. The second question lies outside the scope of this paper, but for the sake of completeness, rather than the addition of anything new, a short explanation needs to be made about phase-out arrangements.

It is probably not possible to formulate a provision for such arrangements which would allow the nature of the phase-out to be ascertained in every case by reference to that provision: the appropriate length of the phase-out period and the nature and size of the annual cuts will vary from one instance to another. Nevertheless, many will hope that, on grounds of equity, the U.N. Law of the Sea Conference will embody in any resolution which it adopts on the subject of fisheries some form of words which will leave nobody in any doubt that coastal states have the obligation to give proper respect to (treat equitably having regard to all the circumstances) the historic rights of foreign fleets in any change of fishery regime in waters not indisputably national on, say, January 1, 1970. Any dispute about such waters or "proper respect" should be referable by any party to the dispute to the International Court of Justice (or to any tribunal nominated for the purpose by the U.N.) with all decisions binding on all parties *save that*, as international tribunals lack the power to enforce their decisions, no state should be obliged to pay any respect to the historic rights claimed by any state which itself refuses to accept the jurisdiction or any decision of such a tribunal.

This may not seem very satisfactory to all foreign fleets, but it is probably as much as they can expect given the inexorable movement to include within national jurisdiction virtually all known fish stocks of any potential value. (A 200-mile limit will leave very little high seas north and east of a line joining the southern tip of Greenland to the southern extremity of the French Atlantic coast.) Indeed, it may be too much to expect given that, in many cases, there is either no basis or an inadequate basis of reciprocity. The knowledge of this may well turn the faces of many states against the idea of creating a general obligation to respect historic rights, though they themselves may be not averse to according such respect in practice. They may prefer to have the freedom of selectivity and to leave phase-out arrangements, if not to bilateral negotiations, then to regional negotiations. This is probably the position of all the major coastal states in the North Atlantic if not elsewhere as well.

CONCLUSION

The preferential fishing system, though not without advantages, must now be regarded as nothing more than a conceptual vehicle which has helped to effect the crossing from the traditional world of high seas fisheries to the island (new found land!) of coastal priority which lies adjacent to the new world of wide-flung coastal jurisdiction. With the destination virtually preordained, there is little point in arguing the relative merits of alternative destinations.

The hope must be, however, that exclusive jurisdiction will not be confused with exclusive use by the coastal state and that fish which could be economically harvested by others will not be allowed to die of old age simply because their capture is beyond the capacity of the coastal state.

In the Northeast Atlantic there are, fortunately, trade-offs which might be used to avoid the worst effects of the discontinuities which might otherwise occur. It would be a pity if this were as much as could be achieved: an economic utilization of resources demands much more than the avoidance of abrupt changes.

In the Northwest Atlantic, the trade-offs appear to be nonexistent: neither the United States nor Canada appears to be interested in fishing in the eastern half of the North Atlantic and neither is likely to be impressed with the argument (always supposing it were demonstrable) that the vessels of other nations could much more economically exploit the fishery resources within North American jurisdiction than could either the United States or Canada. It will be interesting to see what appeals to New World equity will produce in the absence of bargaining counters from other countries (and over the impotent body of economic argument).

Notes to Chapter 7

1. The meaning to be attributed to "stranger" vessels must depend on what the context requires to be understood by "local" vessels. In some cases the latter will refer to a small locality and in others to a whole region and, in both such cases, "strangers" will include vessels from other parts of the same coastal state. In other cases again, the context may require local vessels to embrace all vessels of the coastal state and, consequently, strangers will consist entirely of foreign vessels. But, even here, a distinction may have to be made between those from nearby states—whose vessels may be similar to those of the coastal state itself—and those from distant water states. While the latter are almost always relatively large and are deeply disliked by local fishermen on that account (if not on others also), this dislike may extend to all stranger vessels—including those from other parts of the coastal state in question; the local fishermen's dislike of the one often differs little from that of the other. For this and wider reasons the expression "stranger" vessels is useful.

2. "Review of the Status of Some Heavily Exploited Fish Stocks," FAO Fisheries Circular No. 313.

Chapter Eight

The Case for Coastal State Jurisdiction

Gunnar G. Schram

The following remarks will deal with the subject of coastal state jurisdiction over marine fisheries, with particular reference to the North Atlantic region.

I think we can all agree that the subject which the organizers have selected as this Workshop's main topic is both timely and highly interesting. Problems of fishery jurisdiction and enforcement loom high on the horizon of lawyers, economists, and statesmen at present. One of the most controversial areas in these matters has indeed been the North Atlantic region, where a certain cod war, which shall remain unnamed, and other disputes have had unfortunate and abrasive effects on the relations of nations with long traditions of friendship. This latest flare-up, now fortunately settled by a two-year fisheries treaty, serves to graphically illustrate the complexity and importance of marine jurisdictional problems in the affairs of nations, not least of those bordering on the North Atlantic. The legal uncertainty which has prevailed in matters of fishery jurisdiction since the failure of the 1960 Law of the Sea Conference to establish limits on the ocean has caused a number of disputes and often lack of constructive conservation action in coastal fisheries. These facts were undoubtedly one of the main reasons for the decision of the U.N. General Assembly, in 1970, to take up the law of the sea in its entirety for revision, not only the question of the seabed and the continental shelf. This revision will enter its first, but undoubtedly not the last, phase in Caracas this summer. At this first substantive session one of the most difficult problems will be the regulation of coastal state powers over its adjacent fisheries, both as regards the jurisdictional nature of its rights in this respect and the fixing of the outward boundaries of the jurisdictional zone.

We heard Mr. Austen Laing of the British Trawlers' Federation speak of the possibilities of solving the present jurisdictional impasse by reverting to a system of preferential rights of fishing for the coastal state. Such a system certainly represents a marked advance from the redundant principle of unfettered freedom of fishing beyond a narrow territorial sea. There are many, however, who believe that a limited system of preferential rights will not solve the problem of just allocation and rational management of marine fisheries, any more than the principle of freedom of fishing did in its days of glory. A recognition in law of coastal state jurisdiction over a fairly extensive area beyond its territorial sea would better solve the problems of allocation and conservation, and bring benefits to a great number of developing coastal nations whose fishing industries are still in infancy.

In this context two questions arise which must be answered. Firstly, why is it desirable that the international community recognize such coastal state powers and, secondly, what should their jurisdictional nature be.

The present status of the world's fisheries constitutes the strongest argument for granting the coastal state jurisdiction over its adjacent fisheries. The present world catch of approximately 60 million metric tons can possibly be increased to about 100 million, but only by utilizing new fish species which have not hitherto been used for human consumption. The total resource is therefore severely limited, but called upon to supply a world population that doubles every thirty years, and will total 7 billion in the year 2000. As regards the traditional food-fish stocks, most of them are already fully exploited and will not bear any increase in fishing intensity without rapid destruction. At the same time, the world's fishing nations are competing in building better and ever larger fishing fleets with the technical and economic power to decimate important high seas fisheries in coming years, if competition reigns unfettered on the high seas and each nation can take what it wants.[1]

The present situation will manifestly lead to a disaster in marine fisheries unless a new course is charted. It is submitted that the future productivity of the world's increasingly important fishery resources can only be safeguarded, in the light of past history, by placing the responsibility for their undiminished productivity primarily on the coastal state, and thus bringing an end to the present chaos and lawlessness inherent in the medieval concept of freedom of hunting and fishing for all on the high seas. Reasons of equity and justice also support the thesis of coastal state powers in this respect, as it is the coastal state that is primarily dependent upon its own coastal resources, not states located in faraway geographical areas, with few exceptions. The coastal state has therefore much greater natural interest than others in protecting the marine resources of its coastal areas and also in utilizing such resources for the benefit of its coastal populations—a principle recognized for the first time by the International Court in the *Anglo-Norwegian Fisheries Case* of 1951.

Let us now take a closer look at the various reasons and arguments which give support to recognition of exclusive coastal state jurisdiction by the Law of the Sea Conference.

Protection and Conservation of
Fish-Stocks

What are the results of international management of the fisheries in the North Atlantic during the past decades? The fisheries of both the Northwest and the Northeast Atlantic have been under constant review and management by international fishery commissions and conventions since the end of the war. The main aim of the regulatory instruments under which the commissions were established is "to ensure the conservation of the fish-stocks and the rational exploitation of the fisheries" in the North Atlantic.[2]

Let us briefly examine the record of the thirty-year period of international management in this area in order to ascertain whether such multilateral management of marine resources can reasonably be said to have succeeded or whether other approaches are called for.

Last year FAO's Department of Fisheries issued a survey showing that in the Northeast Atlantic one major fishery was already depleted (the Atlanto-Scandian herring) and two others were fully exploited or depleted (North Sea herring and the North Sea mackerel)[3] In the Northwest Atlantic one fish stock was found depleted (the Georges Bank haddock) and two other fisheries fully exploited and possibly depleted (the Georges Bank silver hake and the herring of the Gulf of Maine and Georges Bank).[4]

Speaking in Ediburgh, Scotland, a few months ago, W. J. Lyon Dean, chairman of Britain's Herring Industry Board, said that the North Sea herrings were perilously close to extinction and only one bad brood year would make this important fishing ground "sterile of herring."[5] Since 1966, attempts have been made within the North-East Atlantic Fisheries Commission to prevent this decline; but they foundered on the lack of unanimity among the members, and consequently the crisis stage described by Dr. Dean was reached last year.[6]

The other example of failure of international fishery regulation in the North Sea is in the mackerel fishery. The catch declined from 0.75 million tons in 1969 to 0.25 million tons in 1971. The stocks declined from over 4 million tons in 1964 to 0.5 million tons in 1970.[7]

As regards the cod-stocks of the Northeast Atlantic, a joint ICES/ICNAF working group on the cod-stocks in the North Atlantic found in its report in 1972 that some of the stocks, such as the Northeast Arctic cod, were overexploited and the fishing effort in the North Atlantic as a whole could be reduced by half without consequent reduction of catches. Only recently have certain protectionary measures for the Arctic cod been put into force, a number of years after a drastic decline in the Barent Sea stocks had taken place.

Another example of the operations of international fishery commissions is a request made by Iceland to NEAFC in 1967 concerning a closure for trawling for cod in an area off the northeast coast. A disagreement among fishery scientists on the scientific basis for implementation of protectionary measures has so far prevented any regulatory action being taken in the area. Such delay, caused by lack of agreement on the interpretation of scientific data, has unfortunately proved to be all too common in the work of the regional fisheries commissions.

In the case of the North Atlantic salmon, the lack of proper conservation action was, however, not caused by a dispute on scientific data, but rather by economic interests of certain members of NEAFC. At the seventh meeting of the Commission, a recommendation on the prohibition of salmon fishery was adopted.[9] This recommendation came to nothing as certain of the members filed objections to it. The question was kept on the agenda of the Commission and at its eleventh meeting, in May 1973, a proposal was adopted prohibiting fishing for salmon in the convention area from January 1, 1976. Three members, however, voted against this proposal and three abstained. And now two members have filed a protest against the fishing ban, so this protectionary action will have little real effect.

I will not elaborate on the failures of international fishery management in the Northwest Atlantic. The various instances are only too well known to the participants in this meeting. Nor is there any need to recapitulate the sad history of the International Whaling Commission. The latest I have seen on that is an Associated Press news item which appeared in the *New York Times* on October 22, 1973: "The United States said today that Japan and the Soviet Union have refused to comply with international decisions for conservation of whales and that their actions constitute a serious setback to protection of the world's whale population."[10] The reason was that Japan had objected to a commission decision to stop taking Antarctic fin whales by June 30, 1976, and that Japan and the Soviet Union had objected to decisions to set a quota for Minke whales to be caught during the next season.

Although fisheries management through international bodies can be both necessary and successful, it is submitted that the cases mentioned show that such management leaves much to be desired with respect to timely conservation action.

But what about the 1958 Convention on Fishing and Conservation of the Living Resources of the High Seas, one might ask in this context. Although this instrument was drawn up only after most careful deliberations in the International Law Commission and at the 1955 Rome Technical Conference, it has failed to solve the problem of high seas fisheries conservation. The reason is a lack of will by individual states to use the machinery of the convention. Only about forty states have ratified it and no state has utilized the provisions of Article VII for implementing conservation action.

Summing up, it seems to be indicated that the most effective way of protecting coastal fisheries is through coastal state regulatory action. Such states obviously have primary interest in preserving the economic advantage comprised by the coastal resources, and experience shows that coastal states are increasingly taking unilateral conservation action in areas even beyond their own jurisdiction.[11] The coastal state does not have to wait for years until protectionary measures can be promulgated, as has been the case in the international commissions. It can act swiftly, as soon as basis for action has been verified.

Utilization and Allocation

Enormous waste takes place in today's world fisheries. FAO's Department of Fisheries estimates that in 1967 a saving of $50-100 million per year was possible in the North Atlantic cod fishery by reducing the amount of fishing. It further estimated that, as the fishing effort has increased considerably there since that time, possible savings are even greater.[12] And, as already mentioned, the joint working group of ICES and ICNAF has estimated that the same catches of cod in the North Atlantic can be achieved with only half the present effort. To a varying degree the same is true of other fisheries around the world.[13] The present unregulated freedom of fishing attracts much more investment in ships, gear, and manpower than is required to land the catch. As long as the yield is economically sustainable for individual nations, often through state subsidies, the fishery is pursued with no thought of overall planning or of the monetary waste it entails for the international community. By granting the coastal state exclusive regulatory powers this waste can be minimized and the economic yield from the fisheries maximized.

In this context it is often said, in opposing coastal state jurisdiction, that entire fish-stocks, which the coastal state is not interested in fishing or does not have the means to exploit, will then be wasted. In actual fact this is only a hypothetical problem, as the coastal state could allow other nations into the fishery, under a licensing system, just as many nations have leased their seabed parcels to foreign companies for oil and gas exploitation.

The main benefit of coastal state jurisdiction is precisely this, *viz.*, that the coastal state can on rational basis reserve a necessary share of the coastal fisheries for its local communities and make the other part available to interested new nations, or nations having traditionally fished in the area. Thus a permanent factor of economic stability is gained for the coastal state, while access for other states would still be kept open.

Enforcement

Even if agreement is arrived at under the present system of international fisheries conventions, there remains the question of enforcement. Usually, enforcement of fisheries regulations on the high seas is the exclusive responsibility of the flag state. Inspectors from other member states are not allowed to control whether, for example, conservation regulations have been fulfilled or not.[14]

Schemes of joint inspection have, however, been adopted under both the North-East and North-West Atlantic Fisheries Conventions, allowing inspectors to board vessels of other contracting parties, but still the flag state remains exclusively responsible for the prosecution of violations.

These inherent weaknesses are not present under the system of coastal state jurisdiction. The coastal state will have the power not only to inspect vessels of its own nationals as well as foreign fishing vessels, but also to prosecute offenders in local courts. This leads to much more effective enforcement of both management and protective fishery regulations.

Pollution Control

Pollution of the oceans is a global problem which needs to be encountered and regulated on a global scale. This is a vast task and will take the international community a long time to accomplish. A good beginning is the 1954 Convention for the Prevention of Pollution of the Sea by Oil,[15] and the 1972 Convention on the Prevention of Marine Pollution by Dumping of Wastes and other Matter, not yet in force.[16] Marine pollution will be one of the many items to be dealt with at the upcoming U.N. Conference on the Law of the Sea. A global solution is, however, not yet in sight and most coastal states will feel the need for adopting their own pollution regulations, hopefully based upon international standards, and will want to acquire the right to enforce both such local standards and international regulations in their coastal waters. Canada has initiated this process by enacting legislation in 1970 on an antipollution zone up to 100 miles from its Arctic coast.[17]

Some negative reaction against this new extension of coastal state jurisdiction could be expected.[18] However, if the jurisdiction of the coastal state in this respect does not interfere unduly with the right of free navigation, is based on internationally accepted standards, and is otherwise adequately circumscribed, it will probably gain acceptance in international law as a new aspect of the principle of self-defense and environmental protection.

The Current Legal Situation

If one surveys the legal situation at present with respect to coastal state jurisdiction it may be seen that over 20 states claim some sort of such jurisdiction beyond the 12-mile limit.[19] This jurisdiction varies from the concept of full territorial sovereignty to an exclusive, and sometimes nonexclusive, fisheries and conservation jurisdiction. The trend towards extended coastal jurisdiction is, however, quite clear from this state practice. It is also common knowledge that a number of states have plans for extending their fisheries jurisdiction beyond the 12-mile limit, but eschewing unilateral action, have decided to wait for the outcome of the Law of the Sea Conference.[20]

The trend toward extended coastal jurisdiction is also evident in the proceedings of the U.N. Seabed Committee. Of the twenty three proposals submitted to Sub-committee II of the Seabed Committee in 1972 and 1973 on jurisdictional rights of coastal states,[21] fifteen envisaged exclusive jurisdiction of the coastal state, while eight propose a preferential rights system or a regional zone on a reciprocal basis.[22]

The majority of the proposals envisage the extension of coastal jurisdiction in the form of an economic zone or patrimonial sea concept under which the coastal state will be allowed to utilize, on an exclusive basis, the resources both of the sea and the seabed and subsoil thereof.[23] It seems likely, therefore, that an arrangement along these lines will be adopted by the conference, either at its Caracas meeting in 1974 or at the third meeting in Vienna in 1975.

What gives an additional support to this view is the fact that the U.N. General Assembly adopted, both in 1972 and 1973, a resolution on permanent sovereignty over natural resources, by over one hundred affirmative votes, where the Assembly reaffirmed

> the inalienable right of States to permanent sovereignty over all their natural resources, on land within their international boundaries, as well as those in the sea-bed and in the subsoil thereof, within their national jurisdiction and in the superjacent waters.[24]

This is nothing less than a declaration of an effective exclusive economic zone, by more than two-thirds of the members of the United Nations.

Looking at the political realities of the current situation, there does not seem, therefore, to be much doubt that an exclusive zone will be introduced shortly as a part of the new law of the sea, either by conference agreement or, should the conference fail, by massive unilateral state action.[25]

Two modalities of manifest importance from the point of view of the international community are here. The first is that the coastal state exclusive jurisdiction extend only to economic resources, but not be allowed to affect freedom of navigation or other traditional freedoms of the sea. The second consideration is that the outer limit of the exclusive economic zone be realistically contained to no more than 200 miles as a maximum, so as not to revert to Selden's *Mare Clausum* theory and give life to Ambassador Pardo's worst nightmares.

Accommodation of Interests of Noncoastal States

The main advantages of coastal state jurisdiction may thus be summed up as follows: (a) As regards conservation of the fish-stocks, it will have markedly beneficial consequences as it makes the coastal state, rather than international fisheries organizations, responsible for taking regulatory measures. (b) The efficiency of fishing operations will be increased, as coastal states can now prevent overcapitalization in the fishery sector and rationalize their fisheries to a much greater degree. (c) Marine resources in their totality will be more rationally exploited and their yield maximized, as the coastal state obviously has an interest in licensing foreign fishing for species not utilized by its own industry. At the same time fishery disputes will become less frequent as the regulatory power now rests solely with the coastal state. (d) The difficult question, for the coastal state, of allocation of the resources will be solved and its economic position thereby strengthened. This is especially important for the developing countries, and the loss it implies for distant water fishing states is comparatively small as these countries are the rich, highly developed ones.

It is therefore submitted that extending coastal state jurisdiction will thus solve many of the problems we are witnessing at present in the world's marine fisheries. But any study of such jurisdiction would be remiss if it neglected to point out that it can also create new problems for noncoastal distant water fleets. Although distant water fishing is nowhere a vital national industry it is, however, true that the allocation of international fisheries to the coastal state can for a period of time cause considerable hardship to localized distant water fishing communities, such as Cuxhaven in the Federal Republic of Germany and Grimsby and Hull in the United Kingdom. But there are several remedies possible for the distant water interests under a coastal state regime. Firstly, there is the possibility of a solution on a regional basis: that states belonging to the same geographical area, especially landlocked and developing ones, be granted fishing rights in neighboring economic zones. This is envisaged *inter alia* in the proposals of Jamaica,[26] China,[27] Afghanistan, and five other states,[28] Uganda and Zambia,[29] and Zaire,[30] submitted to the U.N. Seabed Committee.

Secondly, the coastal state might allow foreign fishermen to fish those stocks which they have traditionally exploited, and are not needed by the coastal state. The foreign fishermen would in all cases be subject to the regulatory power of the coastal state, in such matters as management and conservation. This could be achieved by either bilateral arrangements or an international convention. This decision to grant such fishing rights would in either case be within the discretionary powers of the coastal state.

The third alternative is recognition of the exclusive jurisdiction of the coastal state, subject, however, to a right of other states to innocent fishing. The foreign fishermen would then be allowed to catch that portion of the maximum sustainable yield of a stock not taken by the fishermen of the coastal state.[31] The coastal state could not prohibit such foreign fishing, but might be given the right to impose a levy on such fishing. This is the idea behind the U.S. proposal on fisheries in the Seabed Committee and an earlier Canadian proposal.

Thus there exist various alternatives for giving recognition to the interests of foreign fishing in the economic jurisdiction of the coastal state. This applies especially to developing landlocked and shelf-locked states, although the coastal state would understandably want to regulate such fishing by others and have the final say in the matter as regards access, utilization, and other relevant factors.

The Global Approach

Finally, the question may be asked why the new, emerging law of the sea should give recognition to exclusive jurisdiction of the coastal state, instead of creating a new global organization for the regulation and conservation of all fisheries beyond the territorial sea, possibly granting the coastal state some preferential rights under such a global system. The fisheries commissions might become an integral part of such an organization and their effectiveness and powers greatly augmented. Thus the oceans and their resources would truly become the common heritage of mankind.

The writer agrees that this would be an ideal solution of the urgent problem of regulating the world's marine fisheries. But one must face facts and view the present situation with due realism. The objections against such an undertaking are the following:

a. Given the inherent reluctance of the big powers to accept any kind of compulsory international control or grant powers which might limit their sovereignty to a new international body, the chances for an adoption of such a scheme must be judged minimal.

b. Approximately one-third of all coastal states have already unilaterally claimed varying degrees of exclusive jurisdiction over fisheries beyond the 12-mile territorial sea, and maintained these claims in practice. It is unrealistic to expect these nations to backtrack and renounce these economic advantages.

c. The United Nations have twice, with more than a two-thirds majority vote, declared that the resources of the superjacent waters are under the sovereignty of the coastal state.[32] Similarly meetings of groups of states on entire continents have opted for a 200-mile exclusive economic zone jurisdiction as opposed to the internationalization of marine fisheries.[33]

d. In a world with gradually declining resources, fish are a commodity of rapidly increasing value.[34] They represent an important economic asset, current or potential, not least for the developing nations, many of whom are planning to establish their own fishing industries. It seems therefore unlikely that coastal nations would desist in the future from taking exclusive control of important protein resources off their coasts, especially as they are now being handed the oil and mineral resources of the seabed, out to 200 miles, on a silver platter.

On the other hand, there is a clear need for world marine fisheries organization with the mandate to regulate and manage all ocean fishing of the area beyond coastal state jurisdiction.[35] Even if that jurisdiction will be fixed at 200 miles, the international area will cover more than 60 percent of the oceans and include a number of important fisheries. This is a task which has been neglected in the deliberations of the U.N. Seabed Committee, but is nonetheless one of the most challenging and urgent projects for maintaining the future productivity of the world's oceans.

Notes to Chapter 8

1. For a description of some recent fleet construction see *Fishing News International* 12 (August 1973): 30, 49.
2. Preamble of the North-East Atlantic Fisheries Convention.
3. "Review of the Status of Some Heavily Exploited Fish Stocks," FAO (Rome, 1973), p. 5.
4. *Ibid.*, p. 8.
5. *Fishing News International* 12 (July 1973): 10.
6. The seriousness of this particular case is shown by the fact that in 1970 the stocks were only 10 percent of the 1947 level.
7. *Report of the Tenth Meeting of the North-Atlantic Fisheries Commission.*
8. *Report of the Tenth Meeting of NEAFC*, p. 10.
9. *Report of the Seventh Meeting of the NEAFC*, pp. 19-22.
10. See also "Whaling Nations Resist Call for Moratorium," *Fishing News International* 12 (August 1973): 133-34.
11. Measures taken unilaterally by Norway to save the mackerel-stocks in the North Sea and Iceland's unilateral prohibition of herring fishing by Icelandic vessels in the Northeast Atlantic are but two recent examples.
12. "Conservation Problems with Special Reference to New Technology," FAO Department of Fisheries (Rome, 1972), pp. 2-3.
13. On the economic factors, see F. T. Christy, Jr., and A. Scott, *The Common Wealth in Ocean Fisheries*, (1965), pp. 217-30, and McDougal and Burke, *The Public Order of the Oceans*, (1962), pp. 472 ff.
14. See A. W. Koers, *The Enforcement of Fisheries Agreements on the High Seas: A Comparative Analysis of International State Practice*, Law of the Sea Institute, University of Rhode Island, Occasional Paper No. 6, June 1970.
15. As amended in 1969, 9 *Int. Legal Materials* 1.
16. 11 *Int. Legal Materials*, 6.
17. 9 *Int. Legal Materials*, 543.

18. See L. Henkin, "Arctic Anti-Pollution: Does Canada Make—or Break—International Law," *Am. Journal of Int. Law* 65, 1, pp. 131–36.

19. See "International Boundary Study," Series A. Limits in the Seas. National Claims to Maritime Jurisdictions, No. 36. March 1973. U.S. Department of State.

20. In this group are, *inter alia*, Norway, Canada, and Australia.

21. A list of the proposals submitted to the Sub-committee in 1972 and 1973 is to be found in the committee's final report. Vol. I, G.A. Official Records: Twenty-Eighth Session, Supplement No. 21 (A/9021) pp. 62–66. The texts of the proposals submitted to the Sub-committee in 1973: *ibid.* Vol. III. pp. 1–115.

22. (1) The United States, (2) U.S.S.R., (3) The Netherlands, (4) Jamaica, (5) Japan, (6) Afghanistan, Belgium, Austria, Bolivia, Nepal, and Singapore, (7) Uganda and Zambia, (8) Zaire.

23. See L. D. M. Nelson, "The Patrimonial Sea" *The International and Comparative Law Quarterly* 22, 4 (October 1973): 685–86.

24. Resolutions 3016 (XXVII) and 3171 (XXVIII).

25. For an interesting discussion on the eventuality of failure, see W. T. Burke, "Consequences for Territorial Sea Claims of Failure to Agree at the Next Law of the Sea Conference," in Alexander (ed.), *A New Geneva Conference, Proceedings of the Sixth Annual Conference of the Law of the Sea Institute*, (University of Rhode Island), p. 37 *et seq.*

26. U.N. Seabed Committee: U.N. Doc. A/AC.138/SC.II/L.55.

27. A/AC.138/SC.II/L.34.

28. A/C.138/SC.II/L.39.

29. A/C.138/SC.II/L.41.

30. A/AC.138/SC.II/L.60.

31. See A. W. Koers, *International Regulation of Marine Fisheries*, (London, 1973), pp. 246–48. This approach is advocated by the proposal of Malta in the U.N. Seabed Committee, A/AC.138/SC.II/L.28.

32. *Supra*, p. 8.

33. The Yaouendé Declaration of June 1972: U.N. Doc. A/AC.138/79; the Santo Domingo Declaration of June 1972: U.N. Doc. A/AC.138/80; Declaration of the Algiers Conference of Non-aligned nations, September 1973: U.N. Doc. A/AC.1/L.646.

34. In the last four years the wholesale price of frozen cod-blocks has risen 200 percent on the U.S. market.

35. A. W. Koers, *op. cit.*, pp. 307–17.

Chapter Nine

Balancing of Non-Coastal and Coastal Fishing Rights

Earle E. Seaton

Towards the fishes of the sea, even more so than towards birds of the air and the beasts of the field, man appears in his most atavistic role. Not entirely lacking in his dealings with fishes is the playfulness or curiosity, even if cruel, which leads him to keep certain birds in captivity to enjoy their singing or admire their beauty. Not yet, however, has there developed the creativity and constructiveness which allows certain animal species to be harnessed for work and others to be pitted against one another for his pleasure or gain. With regard to fish species, man reveals himself as a hunter, and often simply a crude and wanton one.

It may be a long leap technologically from the man waiting for hours beside a seashore to snare a fish or two to the fleet of trawlers voyaging thousands of miles during several months to catch, clean, and can some tons of fish, but there is not much change in attitude or behavior towards the booty or the haul. It is still treated as practically inexhaustible, to be accessible to oneself but denied to others.

Fish in the sea are no respecters of persons, property, or national boundaries. Further, they possess incorrigible mobility. Attempts to control or regularize man's relations with them accordingly falter in the no-man's-land of the high seas. In few other branches of jurisprudence does the prevailing situation so closely conform to the primordial law of the jungle. In an attempt to improve this situation, the community of nations is convening for the third time in modern times under United Nations auspices a conference on the law of the sea.

As far as the balancing of fishing rights of noncoastal and coastal states is concerned, it is like balancing the rights of the elephant and the deer to the tender green leaves of the forest trees. The deer, if they could speak, would probably insist on their right to the lowest growing leaves and those closest to their home grazing areas; they would justify their claims with the proposition (if they could speak, it would most likely be in French) "Il faut vivre." The elephants would probably scorn to reply, but if they did, they would most likely confine themselves to: "Pourquoi?"

The fact is that most of the states engaged in distant water fishing are large, powerful, and technologically advanced, while those who confine their fishing to coastal areas are small and underdeveloped. Exceptions occur in this sphere as in the case of Ghana and Republic of Korea, which have developed distant water capabilities. There are also anomalies, as in the case of the United States, which appears to subject its distant water fishing interests to the interests of its navy.

VIEWS OF THE COASTAL FISHING STATES

One view of the rights of coastal fishing states is that of the Latin Americans, as expressed in the draft treaty proposals of Colombia, Mexico, and Venezuela, submitted to the United Nations Committee on Peaceful Uses of the Seabed on April 2, 1973.[1] It is assumed by the proponents that the width of the territorial sea will be 12 miles; within this area the coastal state will have sovereignty, i.e., exclusive rights and control over the mineral as well as living resources therein. The area adjacent to the territorial sea is called "the patrimonial sea"; its outer limits are not to exceed 200 miles; it is proposed that within this area the coastal state should have sovereign rights over the renewable and nonrenewable natural resources which are found in water, the seabed, and the subsoil. The coastal state is to exercise jurisdiction and supervision over exploration and exploitation of resources in the patrimonial sea, but, in exercising such powers, should take "appropriate measures" to ensure that such activities are carried out "with due consideration for other legitimate uses of the sea by other states."

·It will be noted that the proposed legal regimes for the territorial and the patrimonial seas differ in that the coastal state will have "sovereignty" over the former but only "sovereign rights" over natural resources in the latter. The significance of the distinction will obviously be great in connection with navigation, customs, pollution control, and other matters. With respect to fishing, the importance of the distinction lies in the fact that in the exercise of its jurisdiction over the patrimonial sea, necessary provision must be made for use of the area by other states. In other words, distant water fishing fleets cannot be excluded altogether from the patrimonial sea, whereas they may be from the territorial sea.

What minimum and maximum rights of distant water fishing will other states have and who will determine them; These questions are left rather vague by the proponents; the implication is that the coastal state will determine what rights shall be given to other states, because the draft treaty is to be read in conjunction with the Resolution of the Law of the Sea of the O.A.S. Inter-American Juridicial Committee of February 9, 1973, in which it was stated *inter alia* that "The future legal system governing the high seas and the exploitation of their resources should be organized on regional and not on worldwide bases."

If the above presumption is correct, then not merely the coastal state but all of the other states of a particular region would determine what distant water fishing rights might be exercised in the patrimonial seas of coastal states. In this connection it will be borne in mind that Colombia, Mexico, and Venezuela are not proponents of extreme views among the Latin American states with regard to fishing.

As far as the continental shelf is concerned, and its possible overlapping with or inclusion in the patrimonial sea, the three-power draft proposes that should this occur, the legal regime provided for the continental shelf shall apply with respect to the part beyond the patrimonial sea; in that part of the continental shelf covered by the patrimonial sea, the legal regime of the latter shall apply.

It will be recalled that the 1958 Geneva Convention on the Continental Shelf applies *inter alia* to fishing. This convention recognizes that all coastal states have sovereign rights over the continental shelf, regardless of state proclamation or occupation, for purposes of exploring and exploiting the natural resources found thereon. It defines the continental shelf as:

> the seabed and subsoil of submarine areas adjacent to the coast but outside the area of the territorial sea, to a depth of 200 metres, or beyond that limit, to where the depth of the superjacent waters admits of the exploitation of the natural resources of the said areas.

Article 2(4) of the convention explains that natural resources

> consist of the mineral and other non-living resources of the seabed and subsoil together with living organisms belonging to sedentary species, i.e., organisms which, at the harvestable stage, either are immobile on or under the seabed or are unable to move except in constant physical contact with the seabed or the subsoil.

The Continental Shelf Convention—which has been treated by the International Court of Justice "as reflecting, or as crystallizing, received or at least emergent rules of customary international law"—thus recognizes the exclusive rights of the coastal state over sedentary fisheries, such as Ceylon long claimed in regard to pearl oysters and sharks beyond the limits of her territorial sea, and Tunis in regard to sponges. With regard to nonsedentary fisheries, the

coastal state has no sovereign or exclusive rights in the waters above the continental shelf. One may therefore conceive of the three-power draft proposal as
permitting gradually increasing rights to foreign fishermen as the distance from
the coastal state grows, thus:

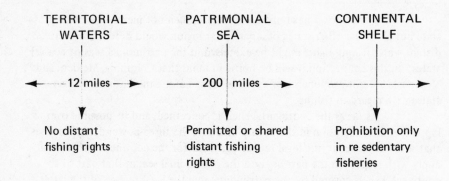

TERRITORIAL WATERS	PATRIMONIAL SEA	CONTINENTAL SHELF
No distant fishing rights	Permitted or shared distant fishing rights	Prohibition only in re sedentary fisheries

What may be considered a less complex—dare I say, less extreme?—
claim has been propounded by Tanzania. Eschewing the more than 100-mile
territorial sea claim of one or two other African States (e.g., Senegal) as well as
the 12-mile territorial sea + 200-mile "economic zone" proposal of Kenya, the
government of Tanzania on August 24, 1973, announced the extension of its
territorial sea limits to 50 miles. On the face of it, this would involve the
assumption of sovereign rights over all the natural resources within the territorial sea, foreign vessels may be completely excluded or permitted to fish only
by license, upon terms that would be fixed at the absolute discretion of the
government of Tanzania.

Beyond the 50-mile territorial sea there would be no further coastal
state rights—no patrimonial sea; nor, in the case of Tanzania, could there be
much exercise of sovereign rights over sedentary fisheries, for the continental
shelf along this part of Africa is little more than 50 miles in breadth. There
would extend beyond the territorial sea only the high seas, in which nationals
of Tanzania would have freedom to fish as well as possible given the present
state of fishing technology.

Who is to say whether such measures as Tanzania's to protect her
coastal fishing are more or less realistic than Kenya's proposals, in very like
circumstances of breadth of continental shelf and development of fishing
technology? Kenya's proposals were submitted to the United Nations Seabed
Committee on August 7, 1972, as draft articles on the concept of an exclusive
"economic zone" beyond the territorial sea.[2] Like the draft treaty of Colombia,
Mexico, and Venezuela, Kenya's proposals assume a 12-mile territorial sea for
all states. Beyond the territorial sea, it is claimed that all states have the right to

establish an economic zone for the primary benefit of their peoples and their respective economies, in which they shall exercise sovereign rights over natural resources for the purpose of exploration and exploitation. The limits of the economic zone are to be determined by the coastal state in accordance with "criteria which take into account their own geographical, geological, biological, ecological, economic and national security factors," but not exceeding 200 miles. These criteria are apparently not to be unilaterally fixed by the coastal state but will be "in accordance with criteria in each region." This implies some kind of machinery for regional consulation, particularly as it is stipulated that disputes arising from delineation of the zone are to be settled in conformity with the United Nations charter and any other relevant regional arrangements. When the economic zone is established, the coastal state shall have "exclusive jurisdiction for the purpose of prevention and control of pollution."

Foreign states will have to obtain permission to exploit the resources of the economic zone, upon terms laid down by the coastal state. Preferential rights of exploiting "the living resources" within the zone shall be given to neighboring states, provided they are developing, landlocked, or with a "small," i.e., narrow continental shelf, and provided the enterprises are "effectively controlled" by the national capital and personnel of such other states. The rights of landlocked or near landlocked states are to be embodied in multi-lateral, regional, or bilateral agreements.

Although the rights of distant water fishing are not completely spelled out in the Kenya draft proposals, it seems clear that here, as in the case of the three-power draft treaty, there is no complete denial of such rights. They exist and may be exercised; but certainly the *modus operandi* and probably the *quantum* of those rights is left to be determined. The Kenya draft proposals seem to reserve to the coastal state alone the determination of the *quantum* and *modus operandi* of distant fishing rights (save in the case of the neighboring states within the same region). However, there seems to be little doubt that the Kenya draft was intended to reflect the spirit of the Conclusions in the General Report of the African States' Regional Seminar on the Law of the Sea, held in Yaourde from June 20 to 30, 1972.[3] Among the recommendations of this Seminar, it was suggested "that African States should promote a new policy of co-operation for the development of fisheries so as to increase their participation in the exploitation of marine resources." This recommendation may be taken to imply that at least vis-à-vis other African states which might possess distant fishing capability, the fixing of terms would be by a process of cooperation.

It will be observed that Kenya's proposed extension of the coastal state's jurisdiction beyond a territorial sea of 12 miles into an economic zone was to be based on geographical, geological, biological, ecological, economic, and national security factors. Presumably the extension of Tanzania's territorial sea from 12 to 50 miles was based on some or all of the same factors. Another instance of an extension of coastal state jurisdiction was Iceland's. On July 14, 1971, the government of Iceland announced its intention to extend its fishing jurisdiction by establishing a zone of exclusive fisheries jurisdiction extending 50 miles from baselines effective not later than September 1, 1972, and also to establish a zone of jurisdiction of 100 nautical miles for protection against pollution.

Previously, between 1958 and 1961, Iceland had extended her fisheries jurisdiction from 4 to 12 miles. The further extension a little over a decade later was necessary, Iceland claimed, in order that the government have control over the fisheries, because the stocks were being overexploited and only national measures could protect them. It was also claimed that the exceptional dependence of Iceland on the fishing industry gave them a special interest, amounting to a right in equity not only to sole responsibility for the management of the fisheries but also to take all the fish above the continental shelf of Iceland (subject to whatever short-term transitional arrangements they might concede). In other words, Iceland's extension of fishery jurisdiction—however unsubstantiated—was based on geographical, geological, biological, ecological, and economic factors.

It may be interesting at this stage to consider the nature and extent of fishing in the areas adjacent to the Icelandic coast. Iceland has very important fisheries resources in its coastal waters—cod, haddock, saithe, and plaice. Within these coastal waters fish most of the Icelandic fishermen, as well as British, Germans, and Belgians; none of Iceland's nationals fish in foreign coastal waters.[4] It would appear that the extension of Iceland's fishery jurisdiction to 50 miles would not carry it beyond the continental shelf. According to current doctrine in international law, Iceland already possesses sovereign rights over the sedentary fisheries on this shelf. The effect of the new policy therefore is to extend the claim of sovereign rights upward through the water column to embrace all fish within 50 miles from baselines. It would appear that Iceland contemplates that if distant fishing is to occur within the 50-mile area, it will only be upon terms to be determined by the Icelandic authorities. Since, again according to current doctrine of international law, a coastal state possesses sovereign right over the mineral or nonliving resources of the continental shelf, the net effect of Iceland's new policy is to establish the equivalent of an "economic zone"—to use Kenyan terminology—save that its area is less than the 200 miles Kenya purposes, and there is no indication of preferential rights for neighboring landlocked states or others within that particular region.

VIEWS OF DISTANT FISHING STATES

It will be interesting at this stage to consider some of the views of the major distant fishing states. One such state is Japan, which submitted to the United Nations Seabed Committee on August 14, 1972, proposals for a regime of fisheries on the high seas.[5]

It will be observed that the Japanese proposals divide coastal states into two categories, according to their degree of development. Those in the category of "developing" will be granted assistance; those in the category of "developed" are presumed to be in no need. What kind of assistance is proffered? "International co-operation...in the field of fisheries and other related industries...with a view to improving the effectiveness of protection of the interests of the developing state." One possible means of assistance would be to provide gunboats and other naval material to police the seas adjacent to a coastal state, in order to detect or deter infringements against fishing regulations laid down by a coastal state to be observed in its patrimonial sea or economic zone. Presumably, however, it is not this kind of assistance that is intended. One notes that according to the Japanese proposal, a developing coastal state will be entitled to preferential rights "to the extent that it is able to catch a major portion of the allowable catch" in its adjacent seas. It seems more reasonable to construe the Japanese offer of assistance as willingness to provide technology, advanced fishing vessels, and equipment as well as expert training personnel, to enable the nationals of developing coastal states to improve their fishing capability. As for the developed coastal state, it would be entitled to an allocation of resources "necessary for the maintenance" of its local industry, i.e., to preserve the *status quo* in area and quantity of fishing.

It is a little puzzling that the Japanese proposal refers to "small-scale coastal fishing" in regard to developed states. This may not necessarily be the case and one wonders, for example, if the coastal fisheries of Iceland and Norway could be so classified.

As to the area or extent of the zone within which preferential rights are to be exercised, the Japanese propose nothing specific; rather, flexibility is urged, "since situations vary greatly according to areas of the sea." The Japanese prefer that "concrete applicable measures implementing preferential rights" should be settled by negotiation and embodied in agreement, presumably bilateral. In any such agreement, one might doubtless envisage a provision for technical assistance as *quid pro quo* for generous recognition of the interests of "traditionally established fisheries of other States." But what would be the consideration for a developed coastal state, such as Iceland, to grant distant fishing rights in its adjacent waters? Goodwill and international comity are perhaps the readiest answers, although one may expect the distant fishing state to consider its refraining from more vigorous fishing activity as a sufficient *quid pro quo*.

The Japanese proposal regarding highly migratory stocks, including anadromous stocks such as salmon, is that they be excluded from all patrimonial seas, economic zones of other exclusive or preferential coastal fishery areas. Anadromous fishes are those which ascend rivers to spawn. It is pointed out in the Japanese draft proposal that the problem of conservation and regulation of anadromous stocks is a limited one affecting a few countries in certain regions and, as such, is already dealt with by existing fishery bodies such as the International Commission for the Northwest Atlantic Fisheries (ICNAF) and the North-East Atlantic Fisheries Commission (NEAFC).

Both of the North Atlantic commissions (ICNAF and NEAFC) have been criticized for their lack of arbitration procedures and of the power to establish national quotas for specific species or areas.[6] These deficiencies, coupled with the general problems all fisheries face, are blamed for having hampered the commissions in their attempt to prevent the overfishing of a number of North Atlantic stocks, particularly cod and haddock. Nevertheless, it is to these two commissions, in the North Atlantic area, that the Japanese propose to leave the conservation and regulation of highly migratory stocks of fish.

With regard to other stocks of fish (but excluding sedentary fisheries since these are covered by the continental shelf doctrine), the Japanese propose certain arbitration procedures in the event of a failure to reach agreement by negotiation between the coastal and other states. The arbitral tribunal would be composed of a body of experts, who would be empowered to give a binding decision. This faith in the judgment of experts is praiseworthy, but would it be shared by the coastal state concerned? As is pointed out by Hale and Wittusen,[7] "Fishery agreements, in general, fail for several reasons: the lack of trust or friendship; competing interests; differing goals; disputes over 'facts' or specific measures." Would any coastal state be prepared to agree to surrender to a body of experts, however distinguished, the powers of decision-making in regard to fisheries in the adjacent waters, particularly if the areas concerned constitute its continental shelf? In this connection, the Japanese proposal may be compared with the measures provided in the 1958 Geneva Convention on Fishing and the Conservation of the Living Resources of the High Seas, particularly in Article 9, regarding procedures for the settlement of disputes: "submission to a special commission of five members, unless the parties agree to seek a solution by another method of peaceful solution, as provided for in Article 33 of the Charter of the United Nations." Article 33 of the charter enumerates these methods as being negotiation, enquiry, mediation, conciliation, arbitration, judicial settlement, resort to regional agencies, or arrangements or other peaceful means of the parties' own choice.

One may now refer to the U.S.S.R. proposals, which were submitted in the form of a Draft Article on Fishing to the United Nations Seabed Committee on July 18, 1972. The Soviet draft—like the Japanese—also divides coastal states into developing and developed. It is assumed that the maximum breadth of any territorial sea or fishery zone will be 12 miles. Beyond that, a developing coastal state would have special rights to take as much of the allowable catch of fish as can be taken by vessels navigating under its flag. What constitutes "the allowable catch" is to be determined by international fishery organizations such as ICNAF and NEAFC, where such exist. Where they do not exist, the coastal state may itself establish regulatory measures in agreement with other states that fish in the area. The regulatory measures must not be discriminatory against fishermen of other states. Coastal states will have power of control over regulatory measures; they may stop and inspect vessels suspected of violations, but investigation and punishment, if any, is to be carried out by the flag state of the suspected vessel. Any disputes arising between the coastal and other states regarding regulatory measures are to be settled by arbitration unless the parties agree to other means provided for in Article 33 of the United Nations Charter.

The difference between developing and developed coastal states in the Soviet draft is that a developed coastal state may reserve to itself only anadromous fish spawning in its rivers. This proposal (which, incidentally, is the opposite of the Japanese) will have the effect of allowing free competition between coastal and distant fishing for nonanadromous stocks, thereby ensuring, presumably, that no potentially valuable foodstuffs are lost through wastage. Some explanation of the reasoning behind the Soviets' proposals is to be found in the Declaration on Principles of Rational Exploration of Ocean Resources adopted in Moscow on July 7, 1972, by the socialist states of Bulgaria, Czechoslovakia, Poland, Hungary, and the U.S.S.R. Calling for more effective scientific research and regulation of fishing on the high seas by international fishery organizations, the five states declare that:

> Existing systems of international regulation of fishing must be continuously improved. The role of regional international fishing organizations should be increased, and their functions broadened; the exchange of scientific, technical and fishery information should be improved with a view to the objective assessment of stocks of fish; and all interested States, without exception, should be given the opportunity to participate in such organizations, on the principle of sovereign equality. It is necessary to give international organizations functions of international verification of compliance with fishing regulations, in view of the fact that such a measure will promote the more effective protection of fishery resources and their maintenance at the maximum sustainable level.

There will be noted in the five-power socialist desire for increased role and functions of regional fishing organizations a certain similarity with the three-power Latin American draft proposals and also, to a certain extent, with the Kenya draft. However, the socialists envisage participation in the regional organizations of "all interested states," i.e., presumably distant fishing as well as regional coastal states. It is doubtful that this is what the authors of the other two drafts had in mind, primarily because of the diversity of size, interests, and capabilities.

On the other hand, the socialists would probably see in such diversity an opportunity to contribute to the regional fishery organizations expertise and other assets which they might otherwise lack. After affirming their support of the struggle of developing countries to establish independent national economies, including fishing, the five powers declare that:

> They will continue to co-operate with the developing countries in the sphere of marine fishing and, to the extent of their own and their partners' capacities, to assist them in the establishment of a modern marine fishing industry with the necessary shore installations, and will broaden their aid in the training of national cadres for fish industries and fishing fleets.

The key to the United States' views on fisheries is the belief that biological and scientific criteria should form the basis for their regulation. In the Revised Draft Article on Fisheries submitted to the United Nations Seabed Committee on August 4, 1972,[8] it is proposed that authority to regulate the living resources of the high seas shall be determined by their biological characteristics and shall be exercised so as to assure their conservation, maximum utilization, and equitable allocation. The United States proposed to divide living resources into three categories: coastal, anadromous, and highly migratory oceanic.

The United States proposes that a coastal state shall regulate and have preferential rights to all coastal living resources to the limits of their migratory range. It shall have similar rights and powers over anadromous resources (e.g., salmon) which spawn in its fresh or estuarine waters throughout their migratory range on the high seas. Whatever its flag vessels can harvest of coastal and anadromous resources may be annually reserved to the coastal state. Insofar as coastal resources are concerned, the United States' proposal would seem to be similar to the Japanese; but the United States is more generous to the coastal state with regard to anadromous resources. Insofar as anadromous resources spawning in a coastal state's rivers are concerned, the United States' proposal is similar to the U.S.S.R.'s, but the United States is less generous to a developing coastal state with regard to other living resources.

What the United States would allow to all coastal states (developing or developed) is equal right to harvest highly migratory resources, which are to be regulated by "appropriate international fishery organizations." Presumably the United States' proposal intends to include regional organizations where relevant. In proposing that any state whose flag vessels "harvest or intend to harvest" a regulated resource should have equal right to participate in such organizations, the United States' proposals go beyond merely ensuring that "due account" is taken of traditionally established fisheries.

The United States proposes that access be provided to other states to harvest such portion of the coastal and anadromous resources as are not utilized by a coastal state's vessels under a system of priorities which would accord privileges in descending order to:

 a. traditional distant fishing;
 b. regional coastal states, particularly landlocked;
 c. all other states, without discrimination.

In this respect the United States is apparently motivated by the same concern as the Japanese that potentially valuable resources not be wasted.

While the United States proposes to give to the coastal state certain powers of search and seizure, ultimate enforcement power rests with the flag state of an offending vessel. In this respect, the United States' proposal is similar to those of the U.S.S.R. and Japan. Elaborate measures for dispute settlement are contained in the United States' proposal, which are to be utilized unless the parties agree to alternatives provided in Article 33 of the United Nations Charter. A five-member arbitral commission is envisaged, composed of specialists in legal, administrative, or scientific questions relating to fisheries. The commission would be empowered to give a binding decision upon any Article.

With reference to dispute settlement within the framework of the global fisheries convention, one can imagine that the majority of disputes regarding fisheries would arise   between nationals of neighboring states or countries not too distant from one another. This would be so because of the desire to reduce fuel and other costs by fishing close to home.

Presumably one would have more confidence in dispute settlement by the courts of one's own region. However, one must also bear in mind the possibility of disputes between coastal and long distant fishing vessels. This might give rise to a situation in which the coastal state engaged in a dispute with a distant fishing vessel would insist that it be judged by an arbitral tribunal composed of other regional states.

In order to gain the trust of the distant fishing vessel, one should probably have as a member of the tribunal an arbitrator from the flag state. But this raises the question of how much participation in the decision-making bodies of a region would be accorded to noncoastal states from outside the region.

DISPUTE SETTLEMENT: A CASE STUDY

The British did not submit any draft of proposals regarding fisheries jurisdiction to the United Nations Seabed Committee. However, their views on the subject may be found in their reaction to the Icelandic resolution about an extension of their fishery limit up to fifty nautical miles from the baselines.[9] The two governments entered into bilateral discussions during which the British sought to persuade the Icelandic government that the proposed extension would have no basis in international law and that if a need for immediate measures to conserve fish stocks around Iceland could be demonstrated, such measures could be discussed with other interested states either within or inside the framework of the appropriate international organization, NEAFC. As an alternative to the extension of Icelandic limits and pending the conclusion of suitable multilateral conservation measures, the British offered to limit their total catch of demersal fish to the average taken by British vessels from the area in the years 1960–1969, i.e., 185,000 metric tons per annum (as against an estimated 208,000 tons for the year 1971). The Icelandic government was not, however, prepared to negotiate further on this basis or to enter into a detailed discussion of the need for conservation.

In these circumstances the British instituted proceedings with the International Court of Justice (ICJ). Following this, there was a second phase of discussions with the objective of reaching an interim arrangement which would not prejudice the views of either side on the substantive issue before the court. During these meetings the British proposed that an interim arrangement should be based on a catch limitation for the British fleet and, in order to give Iceland a degree of preferential treatment, that there should be no corresponding restriction on the catch of Icelandic vessels. It was also proposed by the British to limit the number of effective fishing days by British vessels in the disputed area, and to consider an arrangement of closed and open seas based on an Icelandic counterproposal.

However, Iceland broke off negotiations, published regulations enforcing the 50-mile limit and, despite an order from the ICJ enjoining Iceland not to enforce such a limit against British vessels, they began systematic harassment of British trawlers. In its order of August 17, 1972, the ICJ, by a 14–1 vote, in effect asked Iceland to refrain from enforcing its regulations against British vessels fishing outside the 12-mile zone, and the United Kingdom to ensure that the annual catch by its vessels was not more than 170,000 metric tons of fish.

A third phase of negotiations subsequently took place, which likewise ended without result. The Federal Republic of Germany, which had conducted similar proceedings before the ICJ, proposed on September 15, 1972, that the negotiations with Iceland should be on a tripartite basis. The United Kingdom accepted this proposal, but Iceland did not. It is an essential element of British policy on international fishery questions that proper measures for the regulation of fisheries in areas outside accepted fisheries limits should be worked out by negotiation among all the countries concerned and not imposed by unilateral action. Equally, the British government fully recognizes that where such measures are found to be necessary, an appropriate degree of preference should be shown to the coastal state. This preference should not, however, extend to the complete exclusion of other countries traditionally fishing in the area.

The British have sought to promote discussion of the conservation problem first within NEAFC and then bilaterally with the government of Iceland. On November 13, 1973, an exchange of notes took place constituting an Interim Agreement between the United Kingdom and Iceland relating to fisheries in the disputed area. The terms of that agreement, which will run for two years and are based on an estimated annual British catch of 130,000 tons, provide *inter alia* that the British trawler fleet will be reduced by comparison with the number of vessels notified as fishing in 1971. Further, British trawlers will not fish in conservation areas during specified periods. Icelandic coast guard vessels have the right to stop any British vessel discovered fishing contrary to the terms of the agreement, but the facts must be verified by summoning the nearest British fishery support vessel. If such violation is found to be proven, the trawler will be crossed off the list of those permitted to fish in the waters of the area.

It is too early to discern what particular conclusions may be drawn from the Britain–Iceland dispute but a few tentative observations may be made. It will be noted that the British have a relatively long history of fishing off Iceland, as their first steam trawlers went there in 1891—they may therefore claim to have traditional fishing interests in that area. Britain is the principal distant fishing nation of Northwest Europe; it has as well some foreign vessels fishing off its coasts, i.e., Belgian, French, and German. The fisheries resources in British coastal waters are of moderate value, consisting of whiting, sole, and herring; relatively few commercially valuable fishing grounds exist within 12 miles off shore.

How should the respective rights of the noncoastal and coastal states be balanced? This is the issue which Britain placed before the International Court of Justice. Britain contended that to the extent that Iceland may, as a coastal state especially dependent on coast fisheries for its livelihood or economic development, assert a need to procure the establishment of a special fisheries conservation regime in the waters adjacent to its coast, it can legitimately pursue that objective by collaboration and agreement with the other countries concerned, but not by the unilateral arrogation of exclusive rights within these waters. In denying the court's jurisdiction, Iceland insisted on its competence unilaterally to assert a zone of exclusive fisheries jurisdiction beyond the former 12-mile limit because of exceptional dependence on its fisheries and the principle of conservation of fish stocks.

In the particular instance of Britain and Iceland, there had been a prior dispute between 1958 and 1961 over Iceland's right to extend her fisheries jurisdiction from four to twelve miles. This dispute was settled by an exchange of notes dated March 11, 1961, in which the British government agreed to withdraw their objections to the 12-mile limit, in return for a 3-year phasing-out period during which British ships continued to fish in a part of the 12-mile zone. The agreement also contained the following clause:

> The Icelandic Government will continue to work for the implementation of the Althing Resolution of 5th May, 1959, regarding the extension of fisheries jurisdiction around Iceland, but shall give to the United Kingdom Government six months notice of such extension and, in case of a dispute in relation to such extension, the matter shall at the request of either party, be referred to the International Court of Justice.

It was on the basis of the aforementioned clause in the 1961 Exchange of Notes Agreement that the ICJ held, by 14 votes to 1, that it had
jurisdiction in the matter of the dispute between the United Kingdom and
Iceland. However, in a separate concurring opinion, Judge Fitzmaurice (of
Great Britain) dwelt on matters which may be considered to be of substance,
rather than the mere preliminary question of jurisdiction. The following is an
excerpt from his opinion:[10]

6. The question of fishery conservation was separately dealt with
   by the 1958 Geneva Conservation Convention concluded in
   quent North-East Atlantic Fisheries Convention concluded in
   London on the 24th January, 1959, of which Iceland, the
   Federal Republic and the United Kingdom were all signatories,
   and the object of which according to its preamble, was "to
   ensure the conservation of the fish stocks and the rational exploitation of the fisheries of the North-East Atlantic Ocean and
   adjacent waters,   which are of common concern to them." But
   agreed measures of conservation on the high seas for the preservation of common fisheries in which all have a right to participate,
   is of course a completely different matter from a unilateral
   claim by a coastal State to prevent fishing by foreign vessels
   entirely, or to allow it only at the will and under the control of
   that State...

4. In a zone known as the "contiguous zone," defined by Article 24
   of the 1958 Territorial Sea Convention..., the coastal State was
   allowed "to exercise the control necessary" for certain specified
   purposes which did not include any right of jurisdiction over
   foreign vessels in order to prevent them from fishing there. In
   other parts of the high seas beyond the contiguous zone, the
   coastal State had no rights of jurisdiction or control at all, except
   in re its own vessels generally: and in re foreign vessels, only as
   recognized in the 1958 Geneva High Seas Convention, viz. for the
   suppression of piracy and the slave trade. flag verification in
   certain cases, as part of the process known as "hot pursuit"...

7. Nor did continental shelf doctrine afford any basis for the
   assertion of exclusive fishery claims by a coastal State merely
   on the ground that its continental shelf underlay the waters
   concerned.
   This was made quite clear by the 1958 Geneva Continental
   Shelf Convention and... was reflected later in the I.C.J. Judgment
   in the North Sea Continental Shelf Case (1969). Article V of the
   Convention... stated that the coastal State exercised "sovereign
   rights" over the shelf for the purpose of exploring it and exploiting its natural resources. But the term "natural resources"

was defined in such a way, in re "living organisms," as to cover
only "sedentary species,"—i.e., "organisms which... either are
immobile on or under the seabed or are unable to move except
in constant physical contact with the seabed or subsoil" (Article
24). The very purpose of this definition was to exclude what
were colloquially known as "swimming fish," or fish which,
whether they at all times swam or not, were capable of so doing
(and this of course includes what are known as "demersal"
species—fish which spend a part of their time on or near the
ocean bed but are swimming fish). Clearly therefore the Convention reserved nothing to the coastal State by way of exclusive
fishery rights, except in what might be called in general terms,
sedentary fisheries...

But in a *Dissenting Opinion* Judge Padilla Nervo (of Mexico) expressed an opposing view on matters of substance. He said: "The most essential
asset of coastal States is to be found in the living resources of the sea covering
their continental shelf and in the fishing zone contiguous to their territorial sea."
Judge Padilla Nervo proceeded to point out that in Latin America and in other
areas of the Third World, there were states that had asserted jurisdiction over
waters   adjacent to their coasts in zones extending in some instances to 130,
150, and even 200 miles.

As Oda has pointed out, the problem of international fisheries
cannot be solved solely by legal techniques.[11] Nor can it be solved on mere
biological criteria. In this connection, the interim agreement between the
United Kingdom and Iceland embodied in the exchange of notes of November
13, 1973, appears to be vulnerable to the criticism Hale and Wittusen have voiced
in general concerning fishery agreements based on restrictions on fishing time
and fishing gear.

Limitations on time only generate greater competition in a shorter
time span, inducing a general tendency for firms to enter the fishery
to get their share before the fishery is closed. The net result of time
limitations may be an extremely short fishing season with lower
average per-vessel price of fish and a greater pressure on, and long-
term overcapacity in, processing, storage, and distribution facilities.
Gear restrictions clearly prohibit the most efficient use of available
technology.[12]

If the short-term result of the United Kingdom–Iceland Interim
Agreement may be economic efficiency, the long-term result may well be
reluctant acceptance by the British of Iceland's extended jurisdiction claim.
Such, at any rate, seems to be indicated by the trend of post–World War II
development of the law of the sea. As Choung Il Chee points out, there have

been three prior major fishing disputes of the postwar period, the Anglo-Norwegian (1951), Korea-Japan (1952-1964), and Anglo-Icelandic (1952-1960). In all three, the legal issues involved were basically similar, namely, the issue of the breadth of the territorial sea and the extent of fishery limits. Also among all three disputes, two distinct tendencies may be noted: (1) the extension of the breadth of the territorial sea had been gradually accepted by the countries which had previously objected to such extension beyond the traditional 3-mile limit; (2) the practices of the system of contiguous fishing zone outside the limit of the territorial sea have been recognized by the countries which had previously refused to recognize any extension of fishery jurisdiction outside the traditional 3-mile limit of the territorial sea.[13]

At present, no hope of a peaceful adjustment of coastal versus distant water fishing rights off Iceland's waters lies in the NEAFC, for the European Fisheries Convention, on which it is based, does not regulate fishery jurisdiction. In any event, this commission has been criticized for its lack of power to establish national quotas for specific species or areas, a defect which, *inter alia*, has allegedly hampered the convention in its attempt to prevent the overfishing of a number of North Atlantic stocks, particularly cod and haddock.[14]

A potential method of peaceful adjustment of coastal versus noncoastal rights exists in the 1958 Convention on Fishing and Conservation of the Living Resources of the High Seas. Article 6(1) of that convention provides that a coastal state has a special interest in the maintenance of the productivity of the living resources in any area of the high seas adjacent to its territorial sea. Article 7(1) allows a coastal state to adopt unilateral measures of conservation in such areas provided that negotiations with other states have not led to agreement within six months. However, certain preconditions must be fulfilled: (a) that there is a need for urgent application of conservation measures in the light of existing knowledge of the fishery; (b) that the measures are based on appropriate scientific findings; and (c) that they do not discriminate in form or in fact against foreign fishermen. If the measures are not accepted by the other states concerned, any of the parties may initiate the compulsory settlement procedures contemplated in Article 9. It is considered that a number of states were unhappy that this convention did not give priority of claims to fishery resources of the coastal states. Furthermore, the provisions for compulsory arbitration are thought to be the chief obstacle to a more general acceptance of the convention.

CONCLUSION

It is to be hoped that the third Conference of the Law of the Sea under United Nations auspices, to be held in Caracas in June of 1974, will provide a means of adjustment of the various coastal and noncoastal interests in the spirit of a community of nations. In order to do so, it must be realized that there no longer exists any mechanical rule such as a 3-mile limit which, regardless of the interests involved, should be applicable under all circumstances and for all purposes. On the other hand, there cannot be complete anarchy. If it were true that coastal fishing grounds, owing to their primordial importance for coastal states and owing to the imminent danger of their complete destruction resulting from the employment of piratical techniques by distant nations, can be adequately preserved only by control and exclusive exploitation by the coastal state, reason and justice doubtless would ensure the rights to such control and exploitation.

But the international community also has a valid interest in ensuring that there is no sudden and drastic derangement of the world economy—which might occur if the vested, long-standing rights of distant fishing nations were abruptly curtailed. Care must be taken that important food supplies for mankind are preserved. For this reason, there may be little tolerance for the small nation without adequate technology to exploit fisheries resources in vast areas adjacent to its coast over which it asserts exclusive or preferential rights.

International law cannot be regarded as a static or congealed body of rules. It must be a developing and viable institution to serve a dynamic and changing world. The history of international law is one of response to economic and political pressures. Thus the 3-mile territorial sea limit was developed, and the concept of freedom of the high seas. If the Santiago Declaration of 1952, by Chile, Ecuador, and Peru, claiming a maritime zone extending about 200 miles to sea was a unilateral extension of jurisdiction, may not the same be said of the Truman Proclamations of 1945 claiming sovereign rights to mineral resources of the continental shelf and over certain (sedentary) fish species? It is not unnatural for countries unendowed by nature with generous continental shelves to seek compensation in extensive patrimonial seas of economic zones. To deny that right, while insisting on the exclusive right to exploit rich mineral areas such as the North Sea, may be very human. But it will not make easier the task of balancing noncoastal and coastal rights. In this day and age, one section of the human community in one part of the world cannot reply "Pourquoi?" to the demand of another section for the means of survival.

Notes to Chapter 9

1. A/AC.138/SCII/L.21.
2. A/AC.138/SCII/L.10.
3. A/AC.138/79.
4. Alexander, "Fisheries in North-West Europe," *Yearbook of World Affairs*, 1960, p. 250.
5. A/AC.138/SC.II/L.12
6. Hale and Wittusen, *World Fisheries* (1971), p. 30.
7. *Ibid.*, p. 42.

8. A/AC.138/SC.II/L.9.

9. White Paper: Icelandic No. 1 (1973), Cmd. 5341, London.

10. Fisheries Jurisdiction Case, February 2, 1973.

11. *International Control of Sea Resources* (Leyden, 1963), p. 142.

12. *Ibid.*, p. 23.

13. Choung Il Chee, "National Regulation of Fisheries in International Law" (Ph.D. dissertation, New York University, 1964).

14. Hale and Wittusen, *op. cit.*, p. 30.

Chapter Ten

Economic Consequences of Extending Fisheries Jurisdiction

Paul Adam

While it is by no means definite what new regime for fisheries will emerge from LOS III, a widespread opinion in specialized circles is that the jurisdiction of the coastal states will be extended up to an economic zone of 200 miles. In such a case, there would not be much room for any other solution than the two following:

1. either a ban on foreign vessels, the bordering state keeping for itself the right of reaping and managing the fish resources in its economic zone (which could be done directly or through common ventures with foreigners); or

2. the sharing of the fish resources through the allocation of catch quotas and/or some kind of fishing effort control.

Only the second solution will be examined in this paper, but it is nevertheless worthwhile to note in passing that the two possibilities are not as opposed as might seem. First, in cases where several 200-mile limits overlap, it is difficult to see any practical solution other than common management by the common management by the bordering states. Furthermore, and more generally, the concept of extended coastal jurisdiction does not exclude the sharing of the

---

The opinions expressed in this paper are those of the author; they do not purport to present the views of the organization to which he belongs.

fish resources of the economic zones. For the sake of diversifying their catches and the activities of their fleets, countries may well agree on some kind of trade-offs resulting in the sharing of the fish resources over which they have jurisdiction. It can even be said that the threat of extended coastal jurisdiction can serve the purpose of accelerating the implementation of international management schemes based on the sharing of the fish resources.

The coastal jurisdiction over a wide economic zone should be understood both ways: it would give rights to reap the resources but also a responsibility to preserve the level of the resources. That ICNAF has so spectacularly launched itself for the past two years into an intricate scheme of quotas could be attributed to a possible threat of unilateral measures, which forced a sudden dynamism into a previously cumbersome international cooperation.

For the time being it is impossible to state whether or not the trend in the direction of an extension of coastal jurisdiction will further cooperation in managing international fisheries. But it is only fair to state that the free regime was not making very remarkable progress in that connection.

These general remarks are aimed only at emphasizing that the projected new regime for fisheries, in spite of significantly increasing the rights of coastal states, will not necessarily exclude, and could even accelerate, the building up of international management schemes for the world fisheries. In this perspective, the first practical measures to be examined are the systems of quotas, on which considerable experience is becoming available.

EFFICIENCY OF THE QUOTA SYSTEM

The quota system is not a panacea capable of giving an automatic solution of all problems in the management of fish resources. The whale case has fully demonstrated that the difficulty of arriving at agreements on the level of the quotas, and the tendency to make quotas too high, might delay depletion rather than stop it. The quota system nevertheless has proved its usefulness in a significant number of cases where fish stocks have been maintained at an acceptable level. It is therefore necessary to examine the conditions which must be fulfilled to ensure maximum possible efficiency of this method of regulation.

The most obvious difficulty is to arrive at compromises acceptable to all the countries concerned. Assuming that the scientists have sufficient information on the behavior of the relevant fish stocks and on the recent recruitment to the same stocks, quotas can be determined at levels which can, barring exceptional happenings, maintain the maximum sustainable yield (MSY).

Here two obstacles are encountered:

1. As the accumulated claims of individual countries might well exceed the total recommended by scientists, an easy way out is to adopt overall quotas higher than the recommended level. And that can work. The scientists may have been pessimistic or have lowered their figures with a view to negotiations

which usually result in increased figures. But the reverse can also be the case; and the danger is that nobody can be 100 percent sure of any such forecast. The definite answer is only given a few years later when more damage has been done, and then it can become necessary to impose significantly more drastic restrictions. This is where unilateral decisions of the coastal state(s) or the threat of such decisions can be useful in maintaining reasonable quota levels.

2. The second obstacle is much less apparent because it is hidden behind the clouds of scientific jargon. Everybody agrees that the quotas should not be higher than required to achieve the MSY. In fact it is true that when a fish stock is sufficiently well documented, the marine biologists can assess the level of the MSY with margins of error which can remain within a 10 percent order of magnitude. But the fishing effort corresponding to this MSY cannot be ascertained with a comparable margin of error. Depending upon the youngest age at which the fish is captured (which varies according to the season, the areas, the techniques used, and the development of those techniques), the fishing effort necessary to achieve the MSY can be   affected by changes of 100 percent and more, with very pronounced changes in the total biomass of the stock. Therefore a control by MSY alone can allow excessive development of the fishing effort, diminishing the size of the total stock, and finally causing depletion through a drop in recruitment. Pelagic shoaling stocks may be very sensitive to this danger, and some have provided outstanding examples of depletion (California sardines, Atlanto-Scandia herring, Peruvian anchoveta). Similar examples for demersal stocks are not so spectacular, but noticeable enough (Barents Sea cod, Eastern Atlantic hake). Such danger has led marine biologists to propose at ICNAF the consideration of an "optimum" sustainable yield which is somewhat lower than MSY (no more than 10 percent in average), but which has the advantage of corresponding to a reasonably nonvarying fishing effort. The proposal was rejected by the commission three years ago, but there are hopes that it, or a similar proposal, could now be examined more favorably. The long-term efficiency of the dynamism recently demonstrated by ICNAF launching itself into quota systems could well depend upon the acceptance of such a tool.

The MSY concept has been very useful in proving the necessity of fish stock management, and was practical before attaining overfishing. In the present state of many stocks more or less extensively overfished, it could be more detrimental than useful.

All the above reasoning covers only the situation of single stocks. There exist stocks the exploitation of which is made in isolation (salmon on the British Columbia coast, halibut in the Northeast Pacific, anchoveta off Peru, etc.), but many more stocks are fished in conjunction with other stocks living in the same or other areas. In this respect fisheries are most often mixed, in the sense that the same vessels are conducting different operations according to season, abundance, etc., or in the sense that the stocks are intermingled and fished at the same time by the same boats.

The mobility of many fishing fleets results in a situation where a quota imposed on a given fish stock calls quickly for other quotas on other fish stocks. The origin of overfishing being obviously an excessive fishing effort which should be partly decreased by the imposed quota, vessels eliminated from a fishery because their quota is attained are diverted to other available stocks in some kind of an apparently unavoidable chain reaction.

Multispecies fisheries, apart from posing intricate technical problems (e.g., as regards mesh sizes) are not always amenable to quota systems. There are cases when a quota on a single stock may be sufficient: that is when the by-catches are either unimportant or relatively constant. But, often, there could exist a very intricate and variable dispersion of the catches according to seasons, types of vessels, and markets supplied; a situation in which a quota or a single stock is completely inefficient. Furthermore, when a single species has a dominant share of the total catches, the variability of the by-catches might be such that stocks of some other species not under quota can be endangered, which should lead to a multiquota system. Unfortunately, in a number of areas such multiquota systems would not be practicable; too many species would be involved and the variability of the existing fishing practices would make enforcement very difficult if not impossible. Such drawbacks are leading to proposals for fishing effort control.

FISHING EFFORT CONTROL

The experience in fishing effort control is limited to various national schemes, most often unsatisfactory because they tend to protect outdated fishing techniques which would be bankrupted if more modern and efficient techniques were authorized.

To avoid an entirely hypothetical discussion, it can be useful to refer to a proposal recently made by the United States in ICNAF for fishing effort regulation in the southern part of the convention area (namely areas 5 and 6). The reasons for the proposal are stated in a well-documented memorandum, the main ideas of which were summarized at the end of the above section.

It will here suffice to concentrate on the proposal which aims at allocating to all countries participating in that particular fishery a given number of standard fishing days. For each type of vessel of each country, a ratio would be allocated so as to allow a practical implementation of the scheme. It was also suggested that a further step could be to extend fishing effort control to the whole area, but no precise proposals were made in this respect.

While it cannot be questioned that a fishing effort control might be necessary where a quota system is inefficient, it is only fair to recognize that the United States proposal would not be easy to implement.

It could be accepted that small coastal boats would not be bound by such international limitations (there is already an escape clause of this nature for Norway in the recent three-countries agreement on Arctic cod in the Northeast Atlantic); the justification would be that the control of the small boats is difficult. But the bigger vessels should all be included. Although, on the whole, those vessels are fishing the same species, they are fishing with variable intensities in different periods, and are obtaining widely different results—so much so that the United States' proposal referring to past achievements of different vessel types suggested allocating a ratio of three to one vessel type of one country and a ratio of 6.5 to the same vessel type when it belongs to another country. The discrepancy should not be attributed to overall equivalent differences sufficient to explain the discrepancy. It would nevertheless result in an apparent discrimination and would also create completely illogical situations. The extension of fishing effort control to other grounds inside the ICNAF area could at least partly reverse the above discrepancy but would surely introduce many others.

When it is remembered that nowadays an important part of the big fishing fleets is technically fit to operate in any ocean, the intricacy of the resulting schemes for fishing effort control might render them unmanageable. A more complete examination of such extreme cases cannot be made without considering the economics of fish production which give the reasons for the variations in efficiency.

METHODS OF COST ASSESSMENT IN INTERNATIONAL FISHERIES

The money values, which are the common instruments in economic assessments are completely inadequate when dealing with international fisheries. Already, in national fisheries, such units might be misleading: when comparing big incorporated firms with small enterprises or skipper-owners, a model expressed only in monetary values will not easily absorb the social factors, so important for many fishing communities. It does not mean that social and economic factors are incompatible. In fact, the opposite would be true. While completely different economic structures can be revealed by comparisons expressed in money values, further computations in the same units might be mathematically accurate but practically inadequate. The marginal value of the monetary unit is not the same for the skipper-owner of an open boat 20-feet long and for a firm running a dozen big stern freezer trawlers.

In any case, when dealing with international fisheries, the differences are so important that it is completely impossible to use monetary values in direct comparisons. According to the countries, one finds:

      a. varying geographical conditions for access to the grounds;

      b. expertise and technicalities which are never exactly the same and could be very different indeed;

      c. consumer habits which can widely differ;

      d. general economic structures, credit availability, regulations for crew members and social schemes, etc., which are far from similar even inside countries as close as the EEC member countries.

Any of those elements would be enough to render monetary comparisons useless. British vessels have been bought by Icelanders who sometimes land their catches in British ports: assuming that the boats are run with equal ability and operating on the same grounds, the accounting of such trips from the United Kingdom to the United Kingdom and from Iceland with unloading in the United Kingdom is certainly widely different. To mention only one reason: the crew number is small in the United Kingdom and significantly higher in Iceland, which does not reflect the degree of efficiency of the crew members in the respective countries but a different organization of the work.

The inadequacy of the monetary units does not mean that the different national fisheries cannot be compared; and it does not imply that international fisheries are incoherent.

The comparisons should be made in material units which are mainly as follows:

      a. the distances to the grounds;

      b. the weight of fish caught;

      c. the type and weight of the products which are landed;

      d. the number of vessels according to type: the number of days fished, steaming, etc.;

      e. the crew number.

Such combinations of material units can well give results which are difficult to combine logically in the framework of a single sea area. For example, taking the freezer trawlers above 1,800 G.R.T. in the ICNAF area, one could find the following figures for the year 1971:

Vessel Class	Number	Days Fished	Cod Catches (in tons)	Other Catches (in tons)	Catches per Day (in tons)	Days Fished per Vessel
Portugal	10	1,801	38,813	–	21.6	180
Germany (F.R.)	13	1,093	35,933	–	32.9	146
		779	–	39,059 (herring)	50.1	
U.S.S.R.	232	25,924	110,000	717,713 (all species)	31.7	112

It is obvious that the catch rates of those comparable vessels are different. They are nevertheless logical technically as well as economically.

1. The Portuguese vessels are mainly interested in cod because they have to supply a very important domestic demand for salted cod. They fish for cod all year around.

2. The Germans are interested in both cod and herring; they can therefore choose the best periods for each of these two species and obtain better catch rates.

3. The Soviet vessels looking for all kinds of food fish are, as regards catch rates, in an intermediate position.

An examination of the activities of those or similar vessels in other areas would contribute to a better understanding of the kind of operations conducted by the different fleets. For example, in the extreme north of the Northeast Atlantic, the Soviet vessels had, also in 1971, more than 3,000 days fishing with catch rates of 52.4 tons for polar cod, a species which is not appreciated in the Western European markets. Conversely, the German trawlers sometimes accepted much lower catch rates than in the ICNAF zone on grounds where they could get some redfish, which had become scarce and attracts good prices in German ports.

To put the matter in general terms, it can be said that the logic of the different fishing fleet operations has to be appreciated for the whole of given fleets and in relation with the markets they supply.

The consequence of this fact is a contradiction that hampers the building up of comprehensive and detailed management schemes. Such schemes can only be based on conservation principles and organized by sea areas; furthermore, they must be realistic, i.e., take as a basis the fleet activities as they are usually taking place. But the coherence which has to be achieved in regionally organized management schemes does not necessarily correspond to the economic coherence of the fleet operations. Hence the difficult, if not insoluble, problems of attaching ratios to vessel types for eventual fishing effort control.

Technical efficiency, as it can be determined for each fleet by catch and effort regional statistics, is easy to assess when the right data is at hand. But economic efficiency can only be assessed in a different set-up: expanded to all the grounds where the fleets operate and restricted to the markets supplied.

THE PROBLEM OF BIO-ECONOMIC MANAGEMENT

The opposition between the conservation measures to be taken regionally and the economic assessments to be made globally seems to present a deadlock, especially if an extension of coastal jurisdiction is to be expected. Such an extension of coastal jurisdiction over a rather wide economic zone is likely to exclude any kind of worldwide body having some authority over world fisheries, the only kind of set-up which could be empowered to put forward the required economic assessments. As a consequence, the regional bodies or the coastal states having or taking the responsibility for fisheries management would be confronted with economic problems that they could not adequately cover.

This is why the present doctrine in the regulatory bodies of the North Atlantic is that the economic problems are problems to be considered by the individual countries rather than by the bodies themselves. In other words, these bodies are basing their work on biological assessments which are made regionally, but they are not making regional economic assessments: they only touch upon national economic assessments which are put forward by individual countries and which can be discussed but are never endorsed by the commissions themselves.

A prerequisite, before any kind of practical solution could be envisaged and discussed, is to have a better knowledge of the overall problem, i.e., to have better data than at present on the fleets mobil enough to operate in different sea regions. This would allow a better understanding of the problem involved and of its actual magnitude.

Before having the right data at hand on the world fishing fleets, it may nevertheless be suggested that the problem is more limited than it would appear after having stressed above the opposition between biological and economic problems of world fisheries.

Those fisheries that are subject to competition between coastal nonmobile fleets and mobile fleets coming from more or less distant harbors are well known. It might be that a case by case approach would provide acceptable solutions. In such a perspective, the situation in the North Atlantic is certainly the most important.

THE NORTH ATLANTIC CASE

In setting up the geographical limits of the North Atlantic fisheries, Lewis Alexander has drawn two lines: one from Cape São Roque (Brazil) to Cape Palmas (Liberia), giving the largest North Atlantic including developed and developing bordering countries; and a smaller North Atlantic north of a line from the South of Florida to Gibraltar including only developed countries (which could nevertheless comprise a few less-developed countries and a number of remote areas lagging behind and having few opportunities other than fisheries for economic development).

The more restricted North Atlantic will be considered first. Its fisheries are almost entirely covered by two regulatory commissions, ICNAF and NEAFC. The situations of those two commissions, if they are considered in isolation, could seem opposite.

Douglas Johnston has classified the ICNAF area as an area where conflicts of interests are maximum because the coastal states are taking a part which is not even 50 percent of the total catches in the area; on the contrary, the NEAFC area is "inclusive," i.e., exploited mainly by the coastal states and therefore most of the conflicts of interest could be solved inside the "club."

This opposition is certainly real. A strict application of the principle of coastal jurisdiction could mean that the United States, Canada, and Greenland would exclude all foreign vessels within 200 miles in the ICNAF area. Conversely, the states bordering the North Sea could be inclined to manage it as a "common" property of the seven (Italy being too far away to be interested and Luxembourg being landlocked) plus Norway, and to exclude the foreign vessels which, at present, take only a relatively small part of the total catches. But it is a theoretical proposition. The most recent developments in both areas do not follow such a direction.

In the western part of the Atlantic, Canada and the United States are certainly pushing forward measures which imply restrictions on the long-distance fleets operating inside their 200-mile zone. But they are by no means excluding them. The quotas for the noncoastal states are still very important indeed and in total are significantly above the quotas reserved to the coastal states.

In the eastern part of the Atlantic, the interests are more complex than might appear from the hypothetical image of a "common" North Sea. In the NEAFC area taken as a whole, there are already important agreements as regards grounds around Iceland, the Barents Sea, and waters near the Faroes; they involve countries such as the United Kingdom, Belgium, Germany, Norway, Iceland, and the U.S.S.R. And Denmark, acting for the Faroes, can be considered as outside the "common" North Sea lake as well as inside. It seems therefore probable that the NEAFC area, and even the North Sea, will not be transformed into closed shops. In any case, it would not be in conformity with the balance desirable for covering all the fishery interests of the coastal states.

The wider North Atlantic, going down to Ecuador, allows the consideration of the big conflict of interests between developing and developed countries which is, in fact, the basic conflict behind the law of the sea problem. Furthermore, in this northern part of the mid-Atlantic are located very important tuna fisheries, tuna being among the rare species which can be fished outside continental shelves in the middle of sea areas that could hardly be claimed as economic zones by any bordering state.

It is well known that tuna fisheries have shown in the last ten years or so the most striking example of increasing fishing effort, because it has been in many instances a profitable kind of operation.[1] It also has the characteristic of being dominated, on the market side, by the developed countries, and conducted to a much larger extent by developed countries in waters sometimes far distant from the flag countries. It is also a fishing effort, the techniques of which have become highly sophisticated and can hardly be created from scratch without expertise and capital; hence an incentive towards common ventures.

As a result, tuna fisheries, in the middle Atlantic as well as in other oceans, are likely to pose, in a new regime of the sea fisheries, a major source of disagreement.

The action undertaken by the newly born regulatory body in the Atlantic, ICATT, concerns only, for the time being, the commercial size for some species and a limit for some by-catches, without knowing whether the limit must be calculated in numbers or in weight. Action towards quotas has been postponed, and suggestions for fishing effort control have been vaguely touched upon but not explored in any depth.

In other words, from the management standpoint, the situation is still practically open, apart from the two important facts that a body now exists and that it has already made excellent and fast progress in accumulating more knowledge of the state of the stocks.

The elegant and subtle distinctions made by Austen Laing between preference, priority, and jurisdiction of the coastal state could not be more appropriate. As he states, the semantic differences are erased by an irregular but inexorable drift towards enlarging coastal jurisdiction.

If one tries to guess how such drifting will affect tuna fisheries, it is sure that at present no serious forecasts can be made. It can nevertheless be hoped that practical interim measures could be quickly envisaged at an international level, which would require some generosity from the side of the developed countries. In tuna fisheries, the phasing-out would concern a few developed countries and would have to be done without direct compensations. Conversely, phasing in, which is in the development plans of quite a number of developing countries, could not always be adequately achieved without financial and technical help. Such a situation may lead to common ventures, but there is no other species for which it would be more desirable also to have an international management policy taking into account the interests of the coastal states as well as the international character of the resources. Any interim action towards such an aim would help the law of the sea discussions more than any action in any other fishery.

CONCLUSION

In a not-too-old report (based on figures about two or three years old as of January 1974), OECD concluded that "there is an important and growing distortion between the fishing capacity and the fish resources." Since the drafting of that report, a number of fishing fleets have continued to increase, not to mention the modernization and technical improvements rendering more efficient fleets of which the global GRT has remained stable; and there are more plans for development than for scrapping.

A more rational management of many fish resources would obviously imply the phasing out of part of the existing fishing effort. It is impossible to imagine that such a phasing out could be done without losers or with equivalent losses for all concerned. A narrow discussion, involving only fisheries interests, would therefore lead to endless discussions. The slow-going action of ICNAF and NEAFC until two years ago was entirely due to such a difficulty. But since then ICNAF has made a notable step forward and if on the surface it may seem to lag behind, NEAFC was in the background for the agreements on Arctic cod and Atlanto-Scandia herring, and was not ignored in the dispute between Iceland and the United Kingdom. Such a recent dynamism shows that when the political will exists, multinational agreements can be reached even where profound and intricate differences in outlook exist.

Whatever the new regime of the seas will be, the interim measures for fisheries at present underway will have a very important impact by giving practical experience on how to build up and to run such a new regime. Fisheries are certainly worth the trouble of getting a smooth passage to the new regime of the seas, even at the cost of some phase-outs which could be regionally negotiated and nationally compensated.

The benefits and losses due to a new regime or interim measures can be calculated for individual fishery sectors, and sometimes they could allow bilateral or multilateral negotiations. But there are cases where trade-offs would have to be unilateral. This should be understood not as a reason for neglecting the matter but, on the contrary, for studying it more carefully. To minimize, for all fisheries, the costs of bringing new management measures into force is the least due the fishermen.

Note to Chapter 10

1. A difference is sometimes made between the species that are living on the continental shelves and the more migratory species that wander over wide portions of the high seas. While such an opposition can be justified by many examples, it has to be remembered that there is a great variety of situations between the two extremes. In practice, although the distinction is real, the borderline would be very difficult to draw and many species could be classified into a third intermediate group, the limits of which would be equally difficult to draw.

In addition to this purely biological problem, it should be stressed that the boundaries of the states are also varied.

Therefore it seems extremely difficult to envisage a variation of the jurisdiction systems according to the behavior of the individual species. Even relatively localized species are rarely in the sphere of the extended coastal jurisdiction of one state only. And, more generally, even the sedentary types of sea animals cannot be entirely isolated; they are part of marine ecosystems the limits of which are not fixed and vary according to oceanographic movements which, in any case, are not known well enough to be used as a basis for legal classifications.

A case by case approach of the different fisheries would nevertheless result in singling out some fisheries which are specific partly because of the behavior of the species concerned (e.g., tuna fisheries, whaling).

Chapter Eleven

The Emerging Right of Physical Enforcement of Fisheries Measures Beyond Territorial Limits

Richard B. Bilder

The present prospect is that the Caracas Law of the Sea Conference will couple agreement on a fairly narrow territorial sea—probably 12 miles—with recognition in some form of the right of coastal states to regulate or manage fishing in adjacent areas reaching substantially beyond their territorial waters. The emergence of a regime of extensive coastal state fisheries jurisdiction has been presaged, of course, by the general trend toward the establishment of broader fishery limits during the past several decades, including the extensive claims asserted by many Latin American nations, Iceland, Canada, and certain other states. It is even more strongly suggested by the preparatory negotiations in the U.N. Seabed Committee for the forthcoming Law of the Sea Conference. In the course of these preliminary discussions, a number of coastal states made it clear that they will press for very broad coastal state regulatory authority over adjacent fisheries, ranging in some proposals out to 200 miles.

The proposals made in the Seabed Committee differ in a number of respects. Some appear virtually equivalent to an extended territorial sea. Others would establish an exclusive fishery zone. Others would grant the coastal state limited regulatory or managerial powers for special conservation and other purposes. Still others are cast solely in terms of coastal state preferential rights. In a number of these proposals broad fisheries regulatory authority is embraced in the concept of an extensive coastal state "economic resource" or "patrimoni-

I would like to express my appreciation to the University of Wisconsin Sea Grant Program, which has facilitated my study of international law problems relating to the oceans.

al" zone. However, whatever the precise form in which these proposals eventually coalesce, it seems likely that the conference will accept the concept of broad coastal state fisheries jurisdiction and that the law of the sea that emerges from the conference will substantially change and limit traditional concepts of freedom of fisheries on the high seas.

The pros and cons of this development have been dealt with elsewhere. My purpose here is to raise some questions concerning the narrower issue of enforcement within any such broadened areas of coastal state fisheries authority. Assuming that the Caracas conference recognizes the right of a coastal state to establish rules governing fisheries well beyond its territorial seas, must this coastal state authority also embrace equally broad-ranging enforcement powers—including, for example, the right to stop, board, search, and seize foreign fishing vessels in these extensive zones; the right to bring such vessels into the coastal state's ports; and the right to punish the offense in the coastal state's own courts? Or is it possible and desirable to consider alternative international legal regimes under which the enforcement rights of coastal states in such extensive fisheries zones might be more limited and subject to international safeguards? If so, what should these limitations and safeguards be? More generally, does the probable shift to recognition of broad coastal state fisheries, economic and other jurisdiction suggest a rethinking not only of our traditional regulatory concepts but of our traditional enforcement concepts as well? Finally, should the ultimate decision as to the scope of permissible coastal state enforcement authority in these newly extended zones be determined solely by unilateral action by the coastal states concerned, or possibly left to gradual resolution through the flux of state interactions under this new regime, or should this decision be one expressly considered and arrived at in the law of the sea negotiations themselves?

The enforcement issue seems worth special attention for several reasons. Clearly, coercive action by one state to enforce its regulations against foreign vessels on the high seas raises sensitive political and emotional questions that can easily cause serious international friction. We have seen this most recently in the Latin American "wet war" and the Anglo–Icelandic "cod war." The chances for friction obviously increase as coastal state fisheries authority expands to cover substantial portions of the oceans. Since more of the high seas will now be affected, any perceived abuse of coastal state authority—particularly through coercive measures—may be seen by other states as posing particularly acute dangers to those traditional freedoms of the sea the international community may wish to continue to protect, such as the freedom of navigation and freedom of scientific research. The concept of some form of international constraints on coastal state enforcement powers in these expanded fisheries zones offers one possible avenue through which the impact of any abuses can be moderated, the likelihood of serious frictions decreased, and other states affected reassured. Indeed, the enforcement question is one obvious basis on which the conference could reach a compromise on the fisheries issue. Other states will clearly be more inclined to accept broad coastal state regulatory authority if coastal state enforcement powers are more limited and subject to international safeguards.

In examining the enforcement issue, it is useful to remember that legal concepts and arrangements are simply tools to serve our purposes. Thus, the concepts of sovereignty and jurisdiction are malleable and can be divided, fragmented, or redefined as flexibly as we wish to meet our needs. For example, the law of the sea has long found it useful to distinguish between a broad concept of coastal state sovereignty within territorial seas—conceived as its plenary authority both to prescribe and enforce its rules over the widest range of activities and events in this zone—and the narrower concept of a coastal state's more limited special purpose or functional jurisdiction in a contiguous zone outside of these limits, validly exercised only insofar as it legitimately relates to the permissible purpose.

Again, international law has long distinguished between so-called "prescriptive" jurisdiction—the capacity of a state to make valid rules and regulations, and so-called "enforcement" jurisdiction—the capacity of a state to apply coercive processes to ensure observance of such rules, as by procedures of investigation, arrest, civil and criminal trials, and entry of judgments of fine, imprisonment, or confiscation.[1] For example, there are many situations in which international law recognizes that a state may validly make rules regulating certain conduct (the proper exercise of "prescriptive" jurisdiction), but that it may not have complete authority to arrest or bring before a court persons violating such rules (the proper exercise of "enforcement" jurisdiction).

Finally, the concept of enforcement jurisdiction itself comprises a variety of discrete procedures, which at least theoretically are capable of division as desired among the potentially interested states. The enforcement process includes, for example, such separable elements as the right to establish special reporting or other compliance-checking requirements, the right to investigate to determine whether rules or procedures have been violated; the right to stop and board a vessel to facilitate investigation; the scope of investigation once aboard the vessel; the right to seize or protect evidence of violation; the right to detain or arrest the crew or the vessel; the right to take the crew or the vessel into particular ports; the right to recover the fruits of the violation; and the right to punish the offenders and deter other violations by a broad range of possible sanctions such as fine, imprisonment, or seizure of the vessel or catch. Conceivably, some of these powers could be exercised exclusively by the coastal state concerned; others could be exercised through national enforcement exclusively by the flag state of the fishing vessel in question; others could be exercised through mutual enforcement by either the coastal state or flag state acting jointly or independently; and others could be exercised by some international fisheries body or other authority. In some cases, one state might be given primary authority to exercise certain types of powers, with another state or states having a residual authority to exercise such powers only if they are not utilized by the state having primary authority. Enforcement authority could also be divided in various ways among different states or international organizations depending on the type of interest or stock involved. For example, the coastal state might insist upon broad-ranging enforcement powers to deter foreign fishing in a particular coastal fishery reserved exclusively to its own nationals' use, but be willing to accept a more limited, shared, or even international enforcement authority respecting stocks it shares with other states or respecting migratory species, or with respect to broad conservation regulations. Again, the scope and nature of coastal state enforcement authority, as opposed to that of other states, could conceivably vary in different portions of its fisheries zone, with the coastal state exercising a broader range of authority closer to its coasts than at the further reaches of any extensive fisheries zone.

Obviously, there is a very broad matrix of possible enforcement arrangements from which to choose. The point is not that such distinctions are necessarily either desirable or workable. It is that they do furnish a wide range of options to accommodate conflicting interests and to solve practical and negotiating problems.

With this in mind, it may be useful to review the way the enforcement issue has been dealt with (1) in the traditional law of the sea as reflected in the Geneva Conventions, (2) under the various coastal state contiguous fisheries regimes thus far established, and (3) in certain of the proposals presented to the Seabed Committee preparatory to the Law of the Sea Conference, concluding with some general comments regarding the factors bearing upon the enforcement issue and the options which might be considered in this respect at Caracas. I will not deal here in any detail with the related and important question of enforcement of international fisheries agreements. This subject has been comprehensively discussed in an excellent recent paper by Dr. Koers,[2] to whom I am indebted.

II

The traditional law of the sea, as reflected in the Geneva Conventions,[3] has broadly protected the concept of freedom of the seas by severely limiting the enforcement authority of the coastal state beyond its territorial limits. The pattern of this law is, of course, familiar. A coastal state has very broad prescriptive and enforcement jurisdiction over foreign-flag vessels within its internal or territorial waters, subject to the right of innocent passage defined in Articles 14 to 23 of the Territorial Seas Convention. However, Article 14(5) of that convention provides that "Passage of foreign fishing vessels shall not be considered innocent if they do not observe such laws and regulations as the coastal state may make or publish in order to prevent these vessels from fishing in the territorial sea." As to the high seas beyond the territorial sea, the general rule, as established in Article 6(1) of the High Seas Convention, is that "Ships shall sail under the flag of one State only, and, save in exceptional cases expressly provided for in international treaties or in these articles, ships shall be subject to its [the flag state's] exclusive jurisdiction on the high seas."

The Geneva Conventions provide very limited exceptions to this rule of very restricted coastal state authority over foreign-flag vessels on the high seas beyond coastal state territorial limits. The most significant exception is under Article 24 of Territorial Seas Convention, which provides, in relevant part, that:

1. In a zone of the high seas contiguous to its territorial sea, the coastal state may exercise the control necessary to:

   (a)   prevent infringement of its customs, fiscal, immigration, or sanitary regulations within its territory or territorial sea;

   (b)   punish infringements of the above regulations committed within its territory or territorial sea.

2. The contiguous zone may not extend beyond twelve miles from the baseline from which the breadth of the territorial sea is measured.

The careful wording of Article 24 restricts the enforcement authority granted to the coastal state to the specific purposes mentioned, which involve activities closely related to well-accepted coastal state governmental interests in its territorial sea, and to a narrowly defined contiguous area. A second exception, recognized in Article 23 of the High Seas Convention, permits a coastal state to enforce its authority on the high seas in cases of "hot pursuit" originating within its internal waters, territorial sea, or contiguous zone. A third exception, relating principally to the suppression of piracy and the slave trade is recognized in Article 22 of the High Seas Convention, which provides:

1. Except where acts of interference derive from powers conferred by treaty, a warship which encounters a foreign merchant ship on the high seas is not justified in boarding her unless there is reasonable ground for suspecting:

   (a)   That the ship is engaged in piracy; or
   (b)   That the ship is engaged in the slave trade; or

      (c)   That, though flying a foreign flag or refusing to show its flag, the ship is, in reality, of the same nationality as the warship.

2. In the cases provided for in sub-paragraphs (a), (b) and (c) above, the warship may proceed to verify the ship's right to fly its flag. To this end, it may send a boat under the command of an officer to the suspected ship. If suspicion remains after the documents have been checked, it may proceed to a further examination on board the ship, which must be carried out with all possible consideration.

3. If the suspicions prove to be unfounded, and provided that the ship boarded has not committed any act justifying them, it shall be compensated for any loss or damage that may have been sustained.

The narrow reach of Article 22, and the carefully spelled-out procedures for its implementation, serve again to emphasize the importance that traditional international law places on the principle of freedom of navigation on the high seas and the right of vessels to be free from interference. Other limited exceptions to this principle of noninterference may be argued to be implicit, though not spelled out, in the provisions of the Continental Shelf Convention, which recognizes that the coastal state exercises sovereign rights over the continental shelf for the purpose of exploring it and exploiting its resources, and in the provisions of Article 7 of the Fishing and Conservation Convention, which permits a coastal state, under certain conditions, to adopt unilateral measures of conservation in identifiable areas of the high seas adjacent to its territorial sea, which other states are required either to accept or to submit to special dispute-settlement procedures.

      As is well known, the 1958 Geneva Conference, which negotiated the Geneva Law of the Sea Conventions, failed to reach agreement on either the breadth of the territorial sea or the legality and breadth of contiguous fisheries zones. A second Geneva conference, meeting in 1960, also failed to resolve these questions. As a result, the conventions are silent on both of these issues, and the question of the reach of coastal state fisheries jurisdiction was relegated to a primarily political rather than legal arena, where it has remained. However, it seems fair to say that the general pattern of the Geneva Conventions, and particularly their strong prohibition of enforcement action against foreign-flag ships on the high seas, suggests a bias against any far-reaching coastal state assertion of fisheries enforcement jurisdiction, at least beyond a 12-mile limit.

III

In the years since the failure of the 1958 and 1960 Geneva Conferences to agree on territorial and fisheries limits, a number of coastal states have asserted claims to regulate fishing in extensive areas off of their coasts. These claims have taken varying forms. Some are claims to broader territorial seas, some to exclusive fisheries zones, some to conservation zones, and some are hard to classify precisely. In view of the large number of states which have asserted claims to 12-mile fisheries zones during this period, and general acquiescence in such claims, there is a strong argument that international custom now recognizes the legitimacy of exclusive coastal state fisheries jurisdiction out to at least 12 miles from its coasts. But a number of claims have reached well beyond 12 miles. These expanding fisheries claims have given rise to recurrent international controversy, particularly when, as in the case of Ecuador, Peru, and most recently Iceland, the coastal state has sought to enforce its regulations against foreign fishing vessels in the affected zone.[4]

In general, such coastal state claims to extensive fisheries authority beyond territorial seas appear to include claims of the right to exercise not only prescriptive jurisdiction but full enforcement jurisdiction as well. Whatever accommodations may be made in practice, the coastal states involved have not unilaterally conceded any restrictions on their enforcement powers in this respect. For example, U.S. law expressly authorizes the U.S. Coast Guard to board, inspect, seize, and bring into port for punishment foreign fishing vessels violating the U.S. 12-mile fishery limits, 9 miles beyond U.S. territorial limits; indeed it authorizes such action as to foreign vessels violating U.S. regulations covering the sedentary living resources of the U.S. continental shelf, which in some cases may extend up to 200 miles off its coasts.[5] The U.S. has frequently exercised such authority.[6]

Canadian, Peruvian, Ecuadorian, and Icelandic law appears to confer unrestricted enforcement powers within such proclaimed fisheries or territorial zones. Similarly, Article 5 of the European Fisheries Convention recognizes that, subject to certain limitations, "the coastal state has the power to regulate the fisheries and to enforce such regulations" within the 12-mile limits established by the Convention.[7] Latin American attempts to enforce their broad fisheries limits against foreign fishing vessels form the background of the South Pacific "wet war."[8] The scope of some of these Latin American enforcement claims is perhaps most dramatically illustrated in Peru's 1954 action against the Panamanian-registered, Greek-owned Onassis whaling fleet, first sighted 110 miles off of Peru's coasts. In this incident, Peruvian naval and air units arrested two other vessels 300 miles from its coast, and a fifth vessel 364 miles off of its coast, impounding the vessels pending payment of a three-million dollar fine.[9]

The recent Anglo-Icelandic "cod war" is another example of the friction generated by attempts to enforce fisheries regulations in such extensive zones.[10] It is interesting, however, that in several recent cases where coastal states have negotiated temporary accommodations with other states affected by their extensive fisheries claims, the coastal states have been prepared to accept some limits on their enforcement powers. The recent Anglo-Icelandic Interim Agreement of November 13, 1973,[11] for example, appears to some extent to restrict Iceland's enforcement authority in its recently proclaimed 50-mile fisheries zone, at least with respect to certain British vessels. Paragraph 6 of the agreement provides:

> An agreed list of vessels which may fish in these waters in terms of this interim agreement shall be established. The Icelandic Government will not object to the named vessels fishing around Iceland as long as they comply with the terms of this interim agreement. Should a vessel be discovered fishing contrary to the terms of the agreement, the Icelandic coastguard shall have the right to stop it, but shall summon the nearest British fishery support vessel in order to establish the facts. Any trawler found to have violated the terms of the agreement will be crossed off the list.

The agreement is not clear as to the ultimate scope of Icelandic authority with respect to vessels violating the agreement. However, it is evidently the intent of the parties that the United Kingdom have at least primary responsibility for enforcement.

Another more comprehensive example is the U.S.-Brazilian Agreement Concerning Shrimp, signed May 9, 1972,[12] which seeks to resolve, on an interim basis, certain conflicts between the two countries arising out of Brazil's claim to a 200-mile territorial sea. The two-year agreement limits the number of U.S.-flag shrimp boats which may fish in certain areas off of the coast of Northern Brazil. U.S. vessels are required to register with an agency of the U.S. government and will be granted a permit to fish in the designated area. During the time the U.S. shrimp boats are off the coast of Brazil, they will be subject to the control of Brazilian authorities and may be boarded for inspection. Failure to produce a permit, when fishing for shrimp, will result in the seizure of the ship by the Brazilian authorities. The vessel and its crew will then be taken to the nearest Brazilian port and delivered to U.S. authorities for prosecution under United States law. The U.S. agrees to give to the Brazilian government $200,000 in order to aid that country in its enforcement responsibilities. In addition to this sum, the United States will further compensate the government of Brazil in the amount of $100 for each day a U.S. shrimp vessel is under the control of Brazilian enforcement authorities due to a violation of the Agreement. In testifying concerning the agreement before the U.S. Senate,[13] Ambassador McKernan explained the procedure as follows:

The enforcement function—including patrol, inspection, boarding and searching—is assigned to Brazil, for the reason that Brazil is the only Party which has suitable vessels and officials in the area at all times. In order to facilitate the execution of this function by Brazilian authorities, the United States undertakes to furnish pertinent information for the identification of United States vessels which have been authorized by the United States to fish in the area of the agreement and to have such vessels carry an agreed identification sign. If a duly authorized Brazilian enforcement officer finds a United States vessel in violation of the Agreement, he may seize and detain the vessel, in which case the vessel shall be delivered as soon as practicable to an authorized official of the United States at the nearest port or other mutually agreed place. The United States must inform the Government of Brazil of the disposition which it makes of any such case of a violation of the Agreement.

These examples suggest that the boarding, inspection, and even detention aspects of the enforcement process are less politically sensitive than are adjudication and punishment. Apparently, other states are in some circumstances prepared to concede to the coastal state investigatory and apprehending powers, so long as they retain ultimate control over the application of sanctions. Similarly, coastal states are in some circumstances prepared to forego authority to impose punishment, so long as they have some assurance that flag states will do so, thus deterring further violations of coastal state law.

International fisheries agreements also present an interesting analogy with respect to the possibility of limiting or dividing enforcement jurisdiction. In the case of some of these agreements, as permitted by Article 24 of the High Seas Convention when pursuant to special agreement between the parties affected, government vessels of one state may enforce the provisions of the agreement against fishing vessels of other parties.[14] Typically, however, even in cases where such mutual enforcement is permitted, the priority of the flag state in enforcement is recognized, and the enforcement role of nonflag states is limited to the right to stop, board, and inspect. Dr. Koers comments that"...all fisheries agreements which create mutual enforcement expressly reserve to the flag state the right to prosecute an offending vessel and to impose penalties."[15] Some agreements provide that the flag state must report the disposition of the case to the complaining state. But Dr. Koers also points out that the general approach of the most recent international fisheries agreements is to place enforcement within established coastal state fisheries zones exclusively within the hands of the coastal state.[16] The example of international fisheries agreements again suggests that the prosecution and sanctioning phases of enforcement are generally viewed as considerably more sensitive than stopping, boarding, and investigation.

IV

The various fisheries and economic zone proposals presented to Subcommittee
II of the Seabed Committee in the preparatory work for the Caracas conference
suggests a variety of regimes, differing both as to the scope and purposes of
coastal state authority, and as to the extent to which this authority is subject to
international safeguards. Typically, these proposals provide that freedom of
navigation in the fisheries or economic zones should be respected. Several of
them deal expressly or implicitly with the enforcement issue, and may suggest
the range of options which will confront the Caracas conference.

At one end of the spectrum are proposals which would vest in the
coastal state almost plenary sovereignty over very extensive areas of the adjacent
ocean. Thus, the draft articles on Fisheries in National and International Zones
in Ocean Space submitted by Ecuador, Panama, and Peru provides in Article A
that: "It shall be the responsibility of the coastal state to prescribe legal provisions relating to the management and exploitation of living resources in the
maritime zone under its sovereignty and jurisdiction...",[17] which these countries
have otherwise indicated extends out to 200 miles. Article C permits the coastal
state to establish detailed provisions under which it may, if it wishes, permit
exploitation by nations of other states. The broad character of the coastal
states enforcement jurisdiction is made clear in Articles E and F of the draft:

ARTICLE E

The coastal State may, within the limits of the maritime zone under
its sovereignty and jurisdiction, board and inspect foreign-flag fishing
or hunting vessels; if it finds evidence or indications of a breach of
the legal provisions of the coastal State, it shall proceed to appre-
hend the vessel in question and take it to port for the corresponding
proceedings.

ARTICLE F

Any dispute concerning fishing or hunting activities by foreign-flag
vessels within the zone under the sovereignty and jurisdiction of the
coastal State shall be settled by the competent authorities of the
coastal State.

In "international seas" outside of these zones of national sovereignty and jurisdiction," the coastal state maintains certain preferential rights but its enforcement powers are more severely restricted, as indicated in Articles K and L:

ARTICLE K

States shall ensure that the vessels of their flag comply with the
fishing and hunting regulations applicable in the international seas;

and they shall punish those responsible for any breach that may
come to their notice.

ARTICLE L

Where a State has good reason to believe that vessels of the flag of
another State have violated fishing and hunting regulations applicable
to the international seas, the former State may request the flag
State to take the necessary steps to punish those responsible.

Under Article J, international dispute settlement arrangements are to be applicable only to disputes involving the "international seas," and not to those
involving the "zones of national sovereignty."

Towards the other end of the spectrum are proposals such as those
of Japan and the Soviet Union, which would establish essentially preferential
fisheries rights primarily for the benefit of developing coastal states. As might
be expected, these proposals would more severely restrict coastal state enforcement rights to investigatory activities and possibly arrest; the actual imposition
of sanctions would remain reserved to the flag-state of the vessel concerned.
Thus, paragraph 5 of the Japanese Draft Article on Fishing,[18] dealing with
"Enforcement," provides:

5.1   RIGHT OF CONTROL BY COASTAL STATES

With respect to regulatory measures adopted pursuant to the present
regime, those coastal States which are entitled to preferential rights,
and/or special status with respect to conservation, have the right to
control the fishing activities in their respective adjacent waters. In
the exercise of such right, the coastal States may inspect vessels of
other States and arrest those vessels violating the regulatory measures
adopted.   The arrested vessels shall however be promptly delivered
to the flag States concerned. The coastal States may not refuse the
participation of other States in controlling the operation, including
boarding officials of the other States on the coastal States patrol
vessels at the request of the latter States. Details of control measures
shall be agreed upon among the parties concerned.

5.2   JURISDICTION

(a)   Each State shall make it an offence for its nationals to violate
any regulatory measures adopted pursuant to the present
regime.

(b)   Nationals on board a vessel violating the regulatory measures
in force shall be duly prosecuted by the flag State concerned.

(c)   Reports prepared by the officials of a coastal State on the
offence committed by a vessel of a non-coastal State shall be

fully respected by that non-coastal State, which shall notify
the coastal State of the disposition of the case as soon as
possible.

Comprehensive international dispute settlement procedures are included.

Similarly the Soviet Draft Article on Fishing,[19] after providing that
a developing coastal state may reserve to itself part of the catch in its adjacent
sea, provides:

4. In those of the areas referred to above where fishing regulatory
measures are carried out through international fisheries organizations, such regulatory regime shall remain effective in the future.

Control over the observance of the fishing regulatory measures
in such areas shall continue to be exercised on the basis of the
provisions adopted within the framework of the respective international fisheries organizations.

5. In the areas referred to in this article which are not covered by
the measures specified in paragraph 4, the coastal State may itself
establish fishing regulatory measures on the basis of scientific
findings. Such measures shall be established by the coastal State
in agreement with the States also engaged in fishing in the said
areas.

Regulatory measures shall not discriminate in form or in substance against fishermen of any of those States.

6. The coastal State may itself exercise control over the observance
of the fishing regulatory measures initiated by it under paragraph 5.

In cases where the competent authorities of the coastal State
have sufficient reasons for believing that a foreign vessel engaged
in fishing is violating these measures, they may stop the vessel and
inspect it, and also draw up a statement of the violations. The
consideration of cases which may arise in connection with violations of the said measures by a foreign vessel, as well as the punishment of members of the crew guilty of such violations, shall be
effected by the flag-State of the vessel which has committed the
violation. Such State shall notify the coastal State of the results
of the investigation and of measures taken by it.

7. Disputes between States on matters connected with the application of the provisions of this article may, at the request of one
of the parties to the dispute, be settled by arbitration unless the
parties agree to settle it by another means of pacific settlement
provided for in Article 33 of the United Nations Charter.

In an Explanatory Note, the Soviet Union comments that:

> The coastal State would be entitled in said areas to exercise control over the observance of the fishing rules so established, including the right to stop and inspect a vessel violating such rules. It is understood that in establishing the said rules, coastal States will co-operate with the countries in fishing in those areas.

A somewhat intermediate position is suggested by Article VIII of the U.S. Revised Draft Article on Fisheries,[20] which, in the context of proposing the so-called "species approach," provides:

VIII. ENFORCEMENT

Actions under this paragraph shall be taken in such a manner as to minimize interference with fishing and other activities in the marine environment.

A. Coastal State—the coastal State may inspect and arrest vessels for fishing in violation of its regulations. The coastal State may try and punish vessels for fishing in violation of its regulations, provided that where the state of nationality of a vessel has established procedures for the trial and punishment of violations of coastal State fishing regulations adopted in accordance with this article, an arrested vessel shall be delivered promptly to duly authorized officials of the State of nationality for trial and punishment, who shall notify the coastal State of the disposition of the case within six months.

B. International fisheries organization—Each State party to an international organization shall make it an offence for its flag vessel to violate the regulations adopted by such organization in accordance with this article. Officials authorized by the appropriate international organization, or of any State so authorized by the organization, may inspect and arrest vessels for violating the fishery regulations adopted by such organizations. An arrested vessel shall be promptly delivered to the duly authorized officials of the flag State. Only the flag State of the offending vessel shall have jurisdiction to try the case or impose any penalties regarding the violation of fishery regulations adopted by international organizations pursuant to this article. Such State has the responsibility of notifying the enforcing organization within a period of six months of the disposition of the case.

The U.S. draft article also provides detailed procedures for dispute settlement by an international commission. These dispute-settlement provisions have presumably now been supplemented by the recently submitted U.S. Draft Articles for a chapter on the Settlement of Disputes,[21] which contemplates the establishment of a special international law of the sea tribunal.

Again, it is interesting that even the Japanese, Soviet, and U.S. proposals are prepared to concede inspection and arrest authority to the coastal state, so long as they retain the ultimate authority over sanctions.

V

It is too early to predict what the Caracas conference will decide as to the nature of the enforcement regime in zones of coastal state fisheries authority beyond territorial limits. However, it may be useful to suggest some of the arguments which may be made and some considerations the conference might wish to take into account.

First, the conference will probably treat the enforcement issue as subsidiary to the broader and more basic question of the general nature and scope of coastal state regulatory authority in adjacent fisheries. That is, while there will doubtless be some interaction between the two, decisions concerning the enforcement regime will probably flow from decisions as to the substantive regime, rather than vice versa. For example, if the conference recognizes a very broad zone as subject to coastal state fisheries authority, the greater risk this poses of interference with other freedom of the sea interests could be expected to increase the pressures for some limitations on the coastal state's enforcement powers. Similarly, decisions as to the enforcement will be affected by the purposes for which the coastal state is granted regulatory authority. Thus, recognition of an exclusive coastal state fisheries zone might be expected to carry with it more comprehensive enforcement powers than would recognition of more limited coastal state preferential rights, managerial or conservatory authority.

Second, coastal states will doubtless be reluctant to accept any limitation on their enforcement authority in areas in which their rule-making authority is recognized and will probably maintain that both aspects of jurisdiction should be commensurate. They will probably say that authority to regulate has little meaning unless coupled with authority to impose sanctions on violators. This is the simplest jurisdictional rule and authority. It has the greatest interest in ensuring compliance with the rules it has established, is most likely to have the facts necessary to establish violation, and can most efficiently enforce such rules since it is physically closest to the situs of the violation and the vessel involved. International experience in related contexts, such as the Oil Pollution Convention, suggests that when sanctions are left to the flag state alone, the enforcement system may have only limited effectiveness. If coastal states are granted extensive multipurpose regulatory authority, covering not only fishing but also, for example, seabed resources and pollution, the case for the efficiency of coastal state enforcement is probably strengthened, since enforcement vessels can police many types of regulations at the same time. Moreover, the issue of enforcement authority could well become involved with nationalistic sentiments of a number of coastal states, many of whom view their adjacent seas as part of their national patrimony. They may consequently see any limitation on their enforcement authority as offensive to national dignity. In view of these considerations, coastal states may take the position that they will accept limitations on their enforcement powers only if an alternative regime is established which offers credible guarantees of ensuring compliance with the rules they establish. If the conference recognizes extensive coastal state jurisdiction without specifically limiting enforcement powers, there seems a good chance that coastal states will claim and assert the full range of such powers.

Third, states with long-distance fishing, navigation, research, or other maritime interests, on the other hand, will obviously be equally reluctant to recognize broad coastal state enforcement powers over extensive areas of the high seas. They will fear that coastal states may abuse these enforcement powers to interfere with other states' legitimate interests. At the extreme, the distinction between such special purpose jurisdiction and complete sovereignty may become indistinguishable, with those extensive zones becoming equivalent to territorial waters in all but name—the concern expressed in the concept of so-called "creeping jurisdiction." Additional problems, such as the question of defining and protecting rights of "innocent passage" by foreign fishing vessels through such extensive coastal state zones, and the question of defining the rights of "hot pursuit" outside of such extensive zones, may add to this concern. Thus, these other states might accept broad coastal state enforcement powers only if it is made clear that something less than full sovereignty—some type of managerial responsibility—is involved, and if such authority is accompanied by credible international safeguards or rights of recourse in the event of abuse. They may urge that, if the coastal state role in such expanded zones is essentially only "managerial," its enforcement authority should be closely limited to those powers strictly necessary to its performance of those managerial responsibilities—which can be argued not to include sanctioning powers. Moreover, these states may argue that a regime of national enforcement by the flag state is in practice more likely to provide effective enforcement than is coastal state enforcement. Thus, effective coastal state enforcement in such broad areas may in practice prove extremely difficult and very costly. For example, a recent authoritative estimate suggests that the U.S. Coast Guard apprehends only some 5 percent of the foreign vessels violating the U.S. 12-mile fisheries zone;[22] in a 200-mile zone the enforcement task would be vastly more difficult particularly for nations with long coastlines. They may urge that the flag state, on the other hand, if it is prepared to do so, can exert a broad range of more effective investigatory, reporting, and sanctioning techniques to ensure compliance by its own vessels with applicable regulations. Flag states may be prepared to take on this task if a jurisdictional regime generally satisfactory to them is achieved and if the alternative is a less desirable coastal state enforcement. However, any restrictions on coastal state enforcement powers in otherwise recognized jurisdictional zones are unlikely to come about through post-conference accommodations. If other states wish to restrict coastal state enforcement powers in any respect, they might be well advised to specifically negotiate and expressly include such limitations at the conference, rather than leaving this question to later resolution.

Fourth, there are a wide variety of possible arrangements potentially available to seek to resolve these conflicting concerns of coastal and other states. As previously suggested, enforcement processes are not unitary but can be tailored and flexibly allocated in any way the countries concerned may decide. Some of the possibilities are illustrated in the practices and proposals earlier discussed. In theory, the overriding objective should be to establish a regime capable of adequately protecting the coastal state interests recognized by the conference with minimal interference—particularly coercive interference—with the interests of other states. There may be various ways of best meeting this objective. Clearly, different types of regulations, designed for different objectives, may suggest different types and allocations of enforcement powers. Thus, any treaty rule established might usefully be designed to permit flexibility and choice among possible options.

Some general points may be worth noting in this respect. For example, we have seen that recent experience suggests that states may no longer view visit and inspection as a crucial interference with the freedom of the seas— at least when exercised under circumstances of probable cause. On the other hand, the actual imposition of punishment by a coastal state over foreign flag ships arrested on the high seas clearly continues to raise more acute problems and potentialities for conflict. Also, as Dr. Koers has pointed out, it may be easier to enforce prohibitions on entry into a zone than to enforce regulations concerning permissible levels or kinds of catches for foreign vessels permitted to enter and fish within a zone. In certain circumstances adequate evidence of violation may be obtained by simply observing and photographing a fishing vessel in a prohibited area or using prohibited gear, without the necessity for stopping or boarding arising. In other situations, official landing or other required reports may provide acceptable evidence of compliance or violation. In still other cases, evidence of continuing violation may suggest more stringent "check-in and check-out" procedures for foreign fishing vessels fishing within a particular area, or even the stationing of inspectors aboard the foreign fishing vessels or their support vessels. In some cases the posting of a bond or deposit by the foreign owner of the allegedly violating vessel through coastal state banking facilities may be a viable alternative to actual arrest of a vessel or crew. If flag state government vessels accompany large fishing fleets of that state, they may be capable of providing, on the complaint of the coastal state, the enforcement the coastal state desires. If inspection is considered necessary or desirable, procedures should be devised which cause minimum interference with fishing operations; a saturation inspection of an entire fleet would clearly seem inappropriate. Again, it is open to the countries concerned to work out various arrangements for allocating or sharing costs of enforcement, as illustrated by the U.S.-Brazil Shrimp Agreement, or even to devise procedures for allocating enforcement costs, as a cost of business, to the fishing fleets concerned. In many cases, common sense and ingenuity may suggest practical nonconflicting accommodations more capable of meeting enforcement objectives than any unitary all-inclusive rules. Indeed, it may be useful to make a closer study of the practical aspects of enforcement problems, in the variety of contexts in which they may arise, before deciding on any single solution or allocations of functions.

Fifth, factors of mutual trust and confidence will affect the type of enforcement regime which coastal and other states consider mutually appropriate or acceptable. Thus, other states will be more inclined to concede broad enforcement powers to a coastal state if they trust that it will not abuse such powers. Correspondingly, coastal states will be more inclined to agree to limit their enforcement authority short of the imposition of sanctions if they trust that the flag state will conscientiously police and prosecute alleged violations. With mutual trust, the choice with respect to the allocation of enforcement competence becomes in part simply a question of comparative efficiency of enforcement. This again suggests that enforcement procedures appropriate and acceptable in one area, such as the North Atlantic, may be less appropriate or acceptable in another. It also suggests that any device, such as reporting requirements or impartial dispute settlement procedures, which increase such trust, make any system more workable.

Sixth, whatever the precise enforcement regime adopted regarding any extensive fisheries or economic zones, there is a strong argument to be made for the establishment in some form of institutionalized compulsory international dispute settlement procedures—perhaps, as suggested in the U.S. proposal, taking the form of a special fisheries or a general law of the sea tribunal. Both coastal and flag states may legitimately fear abuse of any self-judging enforcement authority vested in the other. As suggested above, the prospect of effective international dispute settlement may increase mutual confidence and thus make either alternative more acceptable. The possibility of impartial third-party settlement would also reinforce the concept that any international recognition of broader coastal state fisheries authority entails not only rights but also international responsibilities. At the extreme, it is possible to visualize an international fisheries tribunal directly adjudicating cases involving fisheries violations brought before it by complaining governments—a type of international criminal court acting directly upon individuals or private companies. For example, a coastal state might have the right to stop and search an allegedly offending vessel, detaining it until it posts bond with the tribunal. If the tribunal finds violation, it could impose fines which might be used for international fisheries research or development purposes. In the case of improper interference without probable cause, the tribunal might conceivably have power to award damages against the state concerned, in keeping with the principle reflected in Article 22(3) of the High Seas Convention. Once basic jurisdictional issues are settled, many fisheries controversies may depend on factual and technical judgments, best resolved by such an impartial specialist tribunal. This would, of course, be an extremely interesting development from the standpoint of international law more generally.

Seventh, there are, of course, important questions as to the effect of any expanded coastal state fisheries jurisdiction on the existing or any prospective regime of regional, or possibly global, international fishing agreements. One possibility, of course, is that increasing coastal state fisheries authority will substantially diminish the scope and significance of international fisheries arrangements. There will simply be less of the seas and fewer fisheries subject to potential international control. But another possibility is that the international community might see existing international regional fisheries arrangements or conceivably a broader global fisheries organization, as a way of managing and accommodating the problems of broader coastal state jurisdiction in a manner less threatening to general community law of the sea interests. Thus, various suggestions have been made for broadening the responsibilities of regional fisheries organizations. Indeed, these organizations could conceivably be restructured to integrate newly recognized coastal state interests and preferences within a comprehensive regulatory framework, tailored to regional conditions and requirements. Clearly, any willingness to develop international fisheries arrangements in this way might offer interesting new options as regards the enforcement issue. To date, no international organization has been directly responsible for the enforcement of an international fisheries agreement,[23] and there are a variety of political, administrative, and financial considerations which appear to militate against direct international enforcement. It seems unlikely that we will soon see the establishment for any international fisheries police or "sea guard," even with respect to agreements dealing solely with migratory species. But experience with more sophisticated international fisheries arrangements, including various types of mutual enforcement techniques, has been growing, particularly in the North Atlantic.[24] The possibility of giving international regional fisheries organizations some type of supervisory participatory role in enforcement in any expanded coastal state regulatory zones, perhaps associated with a regional fisheries dispute settlement procedure or tribunal, offers another option which the Caracas conference should explore.

Finally, it can be argued that the precise nature of the legal regime emerging from the Caracas conference may, in the long run, be less significant than the extent to which the conference results reflect a broad and realistic consensus among nations. There are various possible arrangements capable of producing a reasonably effective management of the ocean and its resources. If a wide measure of certainty and predictability as to applicable rules can be achieved, nations and various users of the oceans will doubtless work out appropriate accommodations within this framework to meet their particular needs. In contrast, a situation in which a substantial group of states refuses to accept the conference results will create great uncertainty and pose serious and continuing dangers of friction. A flexible approach to the enforcement issue offers one obvious avenue for achieving compromises capable of producing a broad consensus and at least temporarily stabilizing the law of the sea.

Notes to Chapter 11

1. See, Restatement, Second, Foreign Relations Law of the United States (1965), Sec. 6.

2. See A. W. Koers, "The Enforcement of Fisheries Agreements on the High Seas: A Comparative Analysis of International State Practice," Law of the Sea Institute, University of Rhode Island, Occasional Paper No. 6 (June 1970). Dr. Koers also discusses these questions in his book, *International Regulation of Marine Fisheries* (1973), pp. 219–35.

3. The four Geneva Law of the Sea Conventions, all opened for signature April 29, 1958, are the Convention on the High Seas, 2 U.S.T. 2312, T.I.A.S. No. 5200, 450 U.N.T.S. 82; the Convention on the Territorial Sea and the Contiguous Zone, 2 U.S.T. 1606 T.I.A.S. No. 5639, 516 U.N.T.S. 205; the Convention on the Continental Shelf, 1 U.S.T. 471, T.I.A.S. No. 5578, 499 U.N.T.S. 311; and the Convention on Fishing and Conservation of the Living Resources of the High Seas, 1 U.S.T. 138, T.I.A.S. No. 5969, 559 U.N.T.S. 285.

4. For discussions of the disputes concerning the Peruvian and Ecuadorian fisheries limits, see, e.g., Loring, "The United States–Peruvian Fishing Dispute," 23 *Stanford Law Review* 391 (1971); Note, "Seizures of United States Fishing Vessels—the Status of the Wet War," 6 *San Diego Law Review* 428 (1969); Wolff, "Peruvian-U.S. Relations Over Maritime Fishing: 1945-1969," Law of the Sea Institute, University of Rhode Island, Occasional Paper No. 4 (March 1970). On the Anglo-Icelandic Fisheries dispute, see, e.g., Bilder, "The Anglo-Icelandic Fisheries Dispute," 1973 *Wisconsin Law Review* 37 (1973).

5. Pub. Law 88-308, May 20, 1964, 78 STAT. 194, as amended by Pub. Law 90-427, July 26, 1968, 82 STAT. 445 (to include the twelve-mile contiguous fisheries zone established by Pub. L. 89-658, Oct. 1966, 80 STAT. 908). The provisions are codified in Chapter 21 of the Title 16, U.S. Code, 16 U.S.C. Secs. 1081-86. See also, for the Coast Guard's authority to enforce these laws, 14 U.S.C. Sec. 89.

6. See, e.g., the discussion of U.S. enforcement of its fisheries laws against Japanese and Soviet fishing vessels, in Oliver, "Wet War-North Pacific," 8 *San Diego Law Review* 621 (1971).

7. See, e.g., the Icelandic fisheries regulations of 1972, reprinted in 11 *International Legal Materials* 1112 (1972) [hereafter I.L.M.] and the Brazilian Decrees establishing Fishing Zones and Regulating Fishing of 1971, reprinted in 10 I.L.M. 1224 (1971). The European Fisheries Convention of 1964 is reprinted in 3 I.L.M. 470 (1964). Compare also the far-reaching enforcement provisions of the Canadian Arctic Waters Pollution Prevention Act, 18-19 Eliz. 2, C. 47 (Can. 1970), reprinted as an Appendix to Bilder, "The Canadian Arctic Waters Pollution Prevention Act: New Stresses on the Law of the Sea," 69 *Michigan Law Review* 1 (1970).

8. For a discussion of the South Pacific "wet war," see the references cited in note 4, *supra*.

9. See, e.g., the discussion of this incident and judicial decision in the case in *The Olympic Victor*, Judgment of the Port Officer of Paita, Peru, 1954, reprinted in 22 *International Law Reports* 278 (1958).

10. See, e.g., Bilder, note 4, *supra*.

11. Exchange of Notes Constituting an Interim Agreement in the Fisheries Dispute Between the U.K. and Iceland, Reykjavik, 13 Nov. 1973 (U.K. Cmnd. 5484).

12. Agreement between Brazil and U.S. Concerning Shrimp, May 9, 1972, reprinted in 11 I.L.M. 453 (1972).

13. Reprinted in U.S. Senate Exec. Report No. 92-37 on "Agreement with Brazil Concerning Shrimp" (92nd Cong., 2d Sess. Oct. 2, 1972), at p. 9.

14. See generally, the discussion in Koers, Note 2, *supra*, and note 24, *infra*.

15. *Ibid.*, at pp. 18-19.

16. *Ibid.*, at p. 4.

17. Draft Articles on Fisheries in National and International Zones in Ocean Space, submitted by Ecuador, Panama, and Peru to SubCommittee 11 of the U.N. General Assembly Committee on the Peaceful Uses of the Sea-bed and the Ocean Floor Beyond the Limits of National Jurisdiction [hereafter Seabed Committee], Doc. A/AC. 138/SC II/L. 54 of 10 Aug. 1973, reprinted in 12 I.L.M. 1267 (1973). This supplements the Draft Articles for Inclusion in a Convention on the Law of the Sea, submitted by the same three countries, Doc. A/AC. 138/SC II/L. 27, 13 July 1973, reprinted in 12 I.L.M. 1224 (1973), which proposes a 200-mile zone of coastal state sovereignty. Compare also the Draft Articles in Fisheries submitted by Canada, India, Kenya, and Sri Lanka, Doc. A/AC. 138/SC. II/L.38 of 16 July 1973, reprinted in 12 I.L.M. 1239 (1973); the Chinese Working Paper on Sea Area Within the Limits of National Jurisdiction, Doc. A/AC. 138/SC. II/L.34, 16 July, 1973, reprinted in 12 I.L.M. 1231 (1973); and the Draft Articles on Executive Economic Zone proposed by Algeria and

thirteen other African states, Doc. A/AC 138/SC. II/L.40, 16 July, 1973, reprinted in 12 I.L.M. 1246 (1973).

18.   Japanese Proposals for a Regime of Fisheries on the High Seas, Doc. A/AC 138/SC. II/L.12, Aug. 14, 1972, reprinted in 12 I.L.M. 25 (1973).

19.   U.S.S.R. Draft Article on Fishing, Doc. A/AC. 138/SC. II/L.6, July 18, 1972, reprinted in 12 I.L.M. 36 (1973).

20.   U.S. Revised Draft Fisheries Article, Doc. A/AC. 138/SC. II/L. 9, August 4, 1972, reprinted in 12 I.L.M. 42 (1973).

21.   U.S. Draft Article for a chapter on the Settlement of Disputes, Doc. A/AC 138/97, 21 August 1973, reprinted in 12 I.L.M. 1220 (1973).

22.   See Oliver, Note 6, *supra* at p. 633.

23.   The Agreement among Japan, the Netherlands, Norway, the Soviet Union, and the United Kingdom concerning an International Observer Scheme for Factory Ships Engaged in Pelagic Whaling in the Antarctic, signed October 28, 1963, reprinted in 3 I.L.M. 107 (1964), was never formally implemented. However certain of these countries concluded special arrangements concerning observers that were implemented in 1972. While these arrangements were not internationally managed, they provide for reports to be made through the International Whaling Commission.

24.   See, e.g., the following discussion in *The Federal Ocean Program: The Annual Report of the President to the Congress on the Nation's Efforts to Comprehend, Conserve and Use the Sea* (April 1973), at p. 19:

> The first full year of operation of the ICNAF International Inspection Scheme was 1972. Under the scheme enforcement officers of one nation are allowed to board other ICNAF members' fishing vessels on the high seas to check for compliance with ICNAF regulations. Under maritime practice, enforcement against vessels at sea is a prerogative usually reserved to the flag nation. The ICNAF scheme, together with a similar one under the sister Northeast Atlantic Fisheries Commission, both involving 20 nations as diverse as Japan and Iceland, the Soviet Union and Portugal, and Canada and Romania, demonstrate that it is possible for nations to cooperate in mutual policing of activities of common concern. Numerous inspections carried out during the year were without incident, and no major infractions of the ICNAF regulations were reported.

Chapter Twelve

Compulsory Settlement of Fisheries Disputes

Thomas A. Clingan, Jr.

"...a compromise is often a misunderstanding
that is acceptable to both parties."

J. P. Cot

INTRODUCTION

The settlement of international disputes is a subject of distinct complexity. The
application of procedures for dispute settlement is frequently at odds with its
supportive theory. Compulsory dispute settlement is, by its nature, contrary to
one of the most basic and honored traditions of international law. This is well
characterized in the writing of Lauterpacht:

> The function of law is to regulate the conduct of men by reference
> to rules whose formal—as distinguished from their historical—source
> of validity lies, in the last resort, in a precept imposed from outside.
> Within the community of nations, this essential feature of the rule
> of law is constantly put in jeopardy by the conception of the sov-
> ereignty of States which deduces the binding force of international
> law from the will of each individual member of the international
> community. This is the reason why any inquiry of a general charac-
> ter in the field of international law finds itself at the very start
> confronted with the doctrine of sovereignty.[1]

This doctrine, the equality of sovereigns, operates as a permanent restraint on the concept of compulsory settlement of disputes, for if sovereigns cannot in theory be subject to the will of one another, they can no less be subject to the will of a group of sovereigns or to an international tribunal. This is an historical element which must be kept in mind while evaluating the acceptability of options open to nations within the international community. It is a political reality which nations must adapt to or change.

Even so, nations have frequently worked out individual or multilateral arrangements to settle quarrels in advance of the actual disputes. In structuring these agreements for pacific dispute settlement, nations often attempt to characterize the kinds of problems which party nations agree to submit to binding settlement. One cannot say that these attempts have been particularly successful. Occasionally, states have reserved those disputes which are "justiciable" for binding arbitration or other judicial settlement. The problem here is, of course, determining which such disputes are justiciable, and which are not. While this problem is analagous to the question of justiciability in domestic law, the nature of international law makes the definition of the term more difficult where more than one nation is involved. Occasionally the problem is described in terms of questions of law as opposed to political questions, but this characterization is of only limited help.

A look at history may provide better perspective. Article 38 of the Hague Convention for Pacific Settlement of Disputes of 1907 dealt with the identification of justiciable issues as follows:

> In questions of a legal nature, and especially in the interpretation or application of international conventions, arbitration is recognized by the signatory Powers as the most effective, and at the same time the most equitable, means of settling disputes which diplomacy has failed to settle.[2]

The phrase "questions of a legal nature" adds nothing by way of precision, but the modifying clause concerning the interpretation and application of conventions is of some limited value.

The first two paragraphs of Article 13 of the Covenant of the League of Nations handled the difficulty this way:

1. The Members of the League agree that whenever any dispute shall arise between them which they recognize to be suitable for submission to arbitration, and which cannot be satisfactorily settled by diplomacy, they will submit the whole subject-matter to arbitration.

2. Disputes as to the interpretation of a treaty, as to any question of international law, as to the existence of any international

obligation, or as to the extent and nature of the reparation to be made for any such breach, are declared to be among those which are generally suitable for submission to arbitration.[3]

These guidelines are more specific: yet, note that the language is structured to suggest that the parties must themselves recognize the subject matter as being suitable for arbitration before such submission can be effected. From these examples we can see that as the obligation to arbitrate becomes more precise, the responsibility to do so without reservation becomes less clear.

Article 36 of the Statute of the International Court of Justice gives jurisdiction to the court for all cases referred to it by the parties, and, in addition, for matters provided for in the U.N. Charter and treaties in force. But it further declares:

> 2. The states parties to the present Statute may at any time declare that they recognize as compulsory *ipso facto* and without special agreement, in relation to any other state accepting the same obligation, the jurisdiction of the Court in all legal disputes concerning:
>
> a. the interpretation of a treaty;
>
> b. any question of international law;
>
> c. the existence of any fact which, if established, would constitute a breach of an international obligation.
>
> d. the nature of extent of the reparation to be made for the breach of an international obligation.
>
> 3. The declarations referred to above may be made unconditionally or on condition of reciprocity on the part of several or certain states, or for a certain time.[4]

The international community at large seems to treat disputes of a legal nature differently from those which are purely political. The problem, of course, is to differentiate between them. Lauterpacht points out the attempt to do so by concluding that if parties to a dispute ask only for the application or interpretation of existing international law, the dispute is legal in nature, or "justiciable." However, if, instead of setting forth rights under existing international law, the parties profess to disregard an existing rule, and seek to change the law, the dispute is more than likely political—a conflict in national interests not of a justiciable character.[5]

The more one thinks about these matters, the more the interplay between justiciability and sovereignty becomes evident. Matters of high national interest creating new obligations between sovereigns have been left to negotiation between those sovereigns on a plane of equality. Justiciable disputes interpreting the results of those negotiations, determinations of fact bearing upon those interpretations, and matters of reparation arising out of the interpretations have been considered fit subjects for binding adjudication, but only, in many instances, where the parties to the dispute agree to that kind of procedure. For this reason, dispute settlement for fisheries cannot be discussed limiting oneself to binding or compulsory procedures alone. Fisheries disputes are fraught with mixed questions calling into play all of the skill and all of the legal tools available to the respective parties. Thus it will be necessary to depart somewhat from the assigned topic, reviewing and analyzing the traditional legal tools, both compulsory and voluntary in nature, to establish a comparative base from which to depart. The material on international dispute settlement is voluminous. The purpose of this particular paper is not to exhaust the topic, but simply to provide a sufficient background in general theory so that discussions can focus on the advisability of one or another method for any particular aspect of the problems of this workshop. We must, as J. P. Cot advised, be prepared to choose between the Wise Man and the Prince.[6]

TYPES OF INTERNATIONAL DISPUTE SETTLEMENT PROCEDURES

At this point, thought should be given to identifying points of commonality as well as differences between the various traditional methods of international dispute settlement.

Diplomatic Negotiations

Direct negotiation is, of course, still a primary method of first level dispute settlement. Because most such negotiations are conducted in privacy, one can never be sure of the degree to which problems are solved by this method. However, the number of bilateral and small multilateral agreements which are openly published gives an indication that the total number of difficulties handled by direct diplomatic negotiations between states is substantial. Furthermore, present-day importance of negotiation in the international community is reflected in the number of documents which preface any requirement for submission of disputes to a binding mechanism with a requirement that there first be an attempt to settle the issue by diplomacy. Language typical of this form of approach is found in the following excerpt from the treaty between Belgium and Czechoslovakia dated April 23, 1929, Article 1:

> Disputes of every kind which may arise between the High Contracting Parties and which it has not been possible to settle by diplomacy shall be submitted, under the conditions laid down in the present Convention, to settlement by judicial means or arbitration, preceded, according to circumstances, as a compulsory or optional measure, by recourse to the procedure of conciliation.[7]

This device, obviously, raises a question with regard to how long negotiation must continue before recourse can be had to other processes. Accordingly, it is quite common to add a statement that a solution must be reached by negotiation within a reasonable period of time. Another facet of the problem was noted by Sohn. He pointed out that where a group of nations have in common a dispute with one other nation, it would be irresponsible to require each nation to negotiate individually with that nation before it could be found that the requirement of prerequisite negotiation had been satisfied. In such situations, other procedures could be adopted without completely exhausting all administrative remedies through negotiations. Diplomatic negotiations, by their very nature, are noncompulsory, yet they may take on an aura or gloss of binding quality when written in as part of a package which provides for binding procedures where negotiations fail.

Good Offices

When negotiations fail, it is sometimes appropriate for a third state, or a prominent international official, to make an offer of *good offices*. The offer, of course, must be acceptable to parties to a dispute in order to set the procedure into motion. Upon acceptance by the parties, the third state is authorized to attempt to bring the parties together and to assist them in reaching an adequate solution between or among themselves. This can be done in a number of ways. The third state can act as a go-between, a line of communication between the parties, and usually has as its role the re-establishment of direct negotiations between the disputing parties. Once negotiations are resumed, good offices automatically cease unless the parties invite the third state to be present at further diplomatic negotiations. The primary factor that differentiates good offices from other forms of dispute settlement is that the state proffering good offices plays no other role than acting as an intermediary. If that state (or other party) assumes an active role in bringing about a settlement, then the process is converted into mediation.

Mediation

The function of a mediator has been defined as assisting the parties "in the simplest and most direct manner, avoiding formalities and seeking an acceptable solution."[8] Mediation proceedings are normally informal in the sense that they are without a report and are confidential in nature. Mediation normally continues until an agreement is reached, or until the mediator or the parties declare that the solutions proposed by him are not acceptable. In that sense, the advice given by the mediator is not binding upon the parties, and it must live or die based upon its inherent logic and persuasive force. Sohn points out that in practice the functions of good offices and mediation are often confused, and are often exercised simultaneously without clear distinction between the two functions. He notes:

For instance, when the Committee of Good Offices was discussed in the Indonesian Question, the United States representative on the Security Council stated that should "the parties accept the Council's good offices, they could request it to act as mediator or conciliator and to suggest a method of settlement, or ask it to perform any other service they desired. So long as both parties join in making such a request, there is obviously no limit to the services which the Council can perform in facilitating a just and lasting settlement of this dispute.[9]

Sohn concludes:

While the powers of the Security Council and its subsidiary organs can thus be extended by an agreement between the parties to a dispute, it does not seem proper to put all these functions under the much narrower label of "good offices."[10]

While it may not be technically proper to combine the functions of the two devices, Sohn's observation that it is often done is correct, and the combination must have certain attractions that would lead parties to consider the formula as an alternative. Flexibility would appear to be the major of these attractions. Thus it appears that in the Inter-American Treaty on Good Offices and Mediation of 1936, the parties provided first for negotiation, then:

The High Contracting Parties may have recourse to the good offices or mediation of an eminent citizen of any of the other American countries, preferably chosen from a general list made up in accordance with the following Article.[11]

The second article calls for the nomination of persons "most eminent by reason of their high character and juridical learning," which suggests that the parties had mediation more than good offices in mind. Yet the third article again makes reference to both forms of settlement; thus one could conclude that flexibility was the goal to be achieved.

According to H. G. Darwin,[12] the functions of a mediator are broad and flexible, falling into two categories. First, it is his responsibility to conduct interstate negotiations. In other words, in an atmosphere of reticence and perhaps even resentment, he must be prepared to introduce new elements into the discussion and relax tensions where possible. In doing this, he may discuss ideas with parties individually, and may actually advance the thought of one of the parties to the other without disclosing its origin. Any and all devices of persuasion are open to him. Secondly, he may within the course of discussions offer one or more substantive solutions to the dispute. This is a matter of judgment on his part, and, if he has particularly high credentials, he may wish to ponder carefully the question of diluting his potential influence by offering too many proposed solutions. On the other hand, a person of such stature may not wish to risk failure on the success of a single proposal. Thus the personal factor is important to the selection of an appropriate mediator. Often the mediator will be a government, rather than an individual, providing the advantage of a larger staff and greater input to the process. The advantage of an unallied mediating individual is, of course, his flexibility, which may not be evident if the same person operates as an agent of a government. The problem of selection and of *ad hoc* vs. permanent mechanisms is a perpetual one involved in most dispute settlement procedures, and will be discussed later. However, it might be useful to note the following of Darwin's guidelines:

1. Though mediation can only be of assistance where the parties in dispute are fundamentally willing to settle their dispute, it can help towards a settlement.

2. Since flexibility is an essential element of the procedure, only to a limited extent could it be useful to attempt to establish institutions or standard procedures for mediation.

3. Though States will continue to be able to offer mediation, the appointment of individuals as mediators is a useful alternative possibility.

4. Since the choice of a mediator depends on the nature of the dispute and a large number of other factors, including nationality, it is important that there should be a wide field from which mediators may be chosen.

5. In the view of the personal qualities and experience desirable in an individual mediator, he might best be found from among persons with experience in the United Nations or other international organizations, either as a national representative or in the Secretariat.[13]

Commission of Inquiry

Such commissions normally address themselves to ascertainment of facts; assisting parties to disputes by providing a fair and impartial determination of the bases from which a dispute might have arisen. In so doing, commissions may be regarded as playing an important part in the process which, upon occasion, has resulted in elimination of further proceedings. Fact finding is also recognized as an independent procedure under Article 33 of the Charter of the United Nations, which requires parties to a dispute which is likely to endanger peace and security to seek a solution by "negotiation, *inquiry*, mediation, conciliation, arbitration, resort to regional agencies or arrangements, or other peaceful means of their own choice" [emphasis added] .[14]

Inquiry was also specifically provided for by the Hague Convention on Pacific Settlement of Disputes of 1907 for questions of fact. Article 9 provides:

> In disputes of an international nature involving neither honor nor vital interests, and arising from a difference of opinion on points of fact, the contracting powers deem it expedient and desirable that the parties who have not been able to come to an agreement by means of diplomacy, should, as far as circumstances allow, institute an international commission of inquiry, to facilitate a solution of these disputes by elucidating the facts by means of an impartial and conscientious investigation. [15]

The only case submitted to a commission of inquiry under the 1899 Hague Convention, the precursor of the above, was the *North Sea* or *Dogger Bank* case. [16] In October 1904, during the Russo-Japanese War, a squadron of the Russian fleet passing through the North Sea fired on British fishing vessels from Hull, thinking them to be Japanese torpedo boats. One of the trawlers was sunk, and several fishermen were killed or wounded. A commission of inquiry was set up consisting of admirals from France, Austria-Hungary, the United States, Great Britain, and Russia, and it was authorized to inquire into the circumstances surrounding the incident and to decide questions of responsibility and blame. While some lawyers might dispute whether the question of fault is a question of fact, the commission rather avoided that issue by finding that there were no torpedo boats among the fishing vessels, rendering the ultimate question somewhat moot. Eventually the Russians paid compensation.

In another "naval" case, a dispute arose out of the sinking of the Netherlands steamer *Tubantia* in 1916, possibly by a German submarine. [17] A commission of inquiry was established by the parties in 1921 consisting of a former member of the Swiss Federal Council as chairman, and naval officers from Denmark, Sweden, Germany, and the Netherlands, and it was charged with determining the cause of sinking. In this case, the commission concluded that the sinking was the result of "the explosion of a torpedo launched by a German submarine," but it assiduously avoided deciding the issue of fault, suggesting that the *Tubantia* may have struck a floating, inert torpedo previously expended against some other target. Nonetheless, the Germans responded in damages once the facts were ascertained.

Several other cases of the same nature indicate that they tend to involve isolated incidents, so that fact-finding is relatively simple. A comparison between the *Dogger Bank* case and the *Tubantia* indicates another important characteristic of commissions of inquiry. They do not normally derive conclusions from presumptions. That is, in the *Dogger Bank* case, the commission concluded that there were no Japanese torpedo boats in the fleet fired upon, and in the *Tubantia* there was no conclusion of fault; thus, unlike a court, no presumptions of guilt, innocence, or fault are used to fill a gap created by the failure to carry a given burden of proof. The facts are recounted and established for the use of the parties, and nothing more is attempted.

It is of note that many cases handled by commissions of inquiry involved naval matters, and were for the large part resolved by experts experienced in such matters. It is also notable that in all such cases deemed appropriate for submission to commissions of inquiry, findings of fact invariably led to settlement. The authority of the report of a commission of inquiry bears a direct relationship to the original scope of the charge. When the charge is limited solely to fact finding, it is almost impossible for one of the parties to reject the findings without appearing in bad faith. Thus, for example, in the case of the *Tubantia*, Germany could hardly deny the distinguished panel's findings that the *Tubantia* had been sunk by a German torpedo. Where, however, the charge to the commission includes broader goals, the appropriateness of a commission of inquiry to make such findings may be challenged, and the question is inevitably raised whether the question should have been referred to a better-suited mechanism.

Conciliation

In his work on international conciliation, Cot makes reference to Sir John Fischer Williams' observation that with respect to conciliation "No nice definition is called for." Yet Cot observes:

> Conciliation is a concept, rather than an institution. The intervention of third parties in the settlement of an international dispute can assume many forms, and must adapt itself to the nature of each individual case. All that matters is success; the procedure is of little importance. The factor that is essential to ensure success is still the wisdom and experience of those who are called upon to give their opinion; the institutional set-up that enables them to intervene is secondary.[18]

Conciliation is to be distinguished from mediation in that the process involves an independent body rather than a third state or other prominent negotiator. It is different from a commission of inquiry in that in addition to elucidating facts, a conciliation commission makes positive proposals for the settlement of disputes. But unlike arbitration, the parties have no obligation to accept the commission's conclusions as binding. According to M. Rolin, when rapporteur to the Institut de Droit International on international conciliation, conciliation has four advantages:

1. It offers the parties to the dispute information and a knowledge

of the opponent's case which is invaluable.

2. It affords an opportunity to the lawyers and politicians involved in the dispute at a national level to refer the matter to a small body of independent and qualified persons for their objective appraisal of the issues and for proposals for their settlement.

3. It takes full account of the sensitivity, susceptibilities and prestige of governments in that it is easier to accept a third party's solution than that offered by the opponent.

4. It leaves unchallenged the liberty and sovereignty of the parties. There is complete secrecy, no obligation to accept the Commission's proposals, no loss of rights or abandonment of position. A state retains its sovereign control to the last stage of the proceedings.[19]

Oppenheim, on the other hand, finds the procedure suspect. He points out that the lack of obligation of the parties to accept the commission's proposals makes matters uncertain, subject to the whims of neutral members who have no objective criteria at their disposal while developing a solution.[20]

The procedure of conciliation has only rarely been adopted. The most noted of these was the Conciliation Commission for Palestine established by the General Assembly in 1948 "to assist the Governments and authorities concerned to achieve a final settlement of all questions outstanding between them." The commission has not yet completed its task but is viewed as having served some useful purpose.[21]

Sir Hersch Lauterpacht has been one of conciliation's most searching critics. He rejects the concept of conciliation as a valid device on two grounds. First, conciliation is not mandatory, hence states adopting the procedure have by that act undermined judicial procedure while at the same time giving the illusion that they are agreeing to peaceful settlement. Secondly, he contends, conciliation erodes the foundations of the Rule of Law, for if a state can choose between a settlement having the force of law, and one which is nonjuridical, it will normally opt for the procedure which meets its political interests. Hence, by clothing conciliation with the appearances of a juridical procedure, the states involved "sanction[s] the degradation of the Rule of Law by reducing it to the level of a simple political argument."[22]

Cot, however, disagrees, pointing out that:

> Conciliation is indeed a method of either modifying or setting aside the Rule of Law. Recourse is not had to a conciliator for strict enforcement of the law. Nevertheless the solution proposed by the Commission does take into account the Rule of Law and makes the relationship to it clear.[23]

Thus one can view conciliation as the intervention in the settlement of an international dispute by a body having no political authority of its own, but enjoying the confidence of the parties to the dispute and entrusted with the task of investigating every aspect of the dispute and proposing a nonbinding solution. Because a conciliation commission rests its effectiveness, if any, solely on the trust and confidence of the parties, it is better suited to resolving specific conflicts than it is to relaxing long-standing tensions involving a number of incidents, or traditional mistrust. Problems of the constitution of commissions will be deferred until later.

Arbitration

Arbitration and juridical tribunals have been left for treatment until last because these are the two methods which truly raise the question of sovereign rights. Modern arbitration has changed little from the form in which it emerged at the end of the nineteenth century. Article 15 of the Hague Convention of 1899 offers the following definition:

> International arbitration has for its object the settlement of differences between States by judges of their own choice, and on the basis of a respect for law.[24]

The 1907 version adds the following: "Recourse to arbitration implies an engagement to submit in good faith to the award,"[25] Arbitration, as envisaged by the Hague Convention, is seen as the most effective and most equitable way of dealing with questions of a legal nature, particularly the interpretation or application of treaties.

Arbitration is preserved for use within the United Nations structure by Article 33,[26] insofar as disputes the continuance of which is likely to endanger the maintenance of international peace and security are concerned. It is listed as one of the devices that shall be utilized before resort to the Security Council under the terms of Article 37.[27] If the order in which the devices are listed is indicative of their nature, it would be of interest that arbitration is listed between conciliation and referral to judicial settlement. The distinction between conciliation and arbitration is clearer than the distinction between arbitration and judicial settlement. In fact, arbitration is a judicial process, and as indicated by the Hague Convention, should be conducted by "judges." However, one distinction between them has been noted by Sohn:

> It is generally agreed that there is at least one difference, formal or structural rather than substantive, between international arbitral tribunals and international courts. Traditionally, that difference lies not in their respective functions but in the method of selection

of the arbitrators and judges. While the parties usually have free
choice with respect to membership of an arbitral tribunal, they
have to adopt a court as it is and can vary its membership only to
a slight degree.[28]

Of course this distinction becomes even less in importance in the case where
arbitration is conducted under a permanent framework such as the Permanent
Court of Arbitration rather than on an *ad hoc* basis.[29]

Furthermore, there does in fact seem to be some difference between
cases handled by the International Court of Justice and arbitral tribunals. This
division tends to fall along functional lines. While any case might be referred to
arbitration without reference to whether it is or is not arbitrable, so long as the
parties agree, the converse is not usually true for judicial tribunals to which
only "legal questions" are normally referred. Thus the only time one may expect a general dispute over the jurisdiction of an arbitral tribunal is in those
cases where a governing treaty specifies certain categories of disputes that are
agreed to be the subject of arbitration. Time and space do not permit an
examination here of the many cases that would shed light on the distinction
between arbitration and judicial tribunals. Let it suffice for the present discussion to say that while judicial tribunals are limited to questions of legal interpretation and the application of legal norms, panels of arbitrators have been known
to consider not only such disputes but disputes of a nonlegal character, provided
that the nonlegal disputes do not encroach on the vital interest or national
honor, or affect the territorial integrity of one or both of the parties. In looking
over the cases referred to arbitration outside of the commercial area, however,
there seems to be a growing tendency to reserve to courts all questions of legal
interpretation (but not claims to change existing law), and retain nonlegal disputes to arbitration after an attempt has first been made to utilize conciliation.
As noted by Hazel Fox, "[many] factors have contributed to the decline in the
use of *ad hoc* arbitration as a method of settling disputes today. Its proper
function today—and it is well to recognize this frankly without further analysis
of defects—must be as a useful residual process where no special tribunal or
separate procedure exists."[30] Thus arbitration seems a natural as a compulsory
device demanding more flexibility than can normally be afforded by courts, both
in the selection of judges and in the type of mechanisms and subjects of dispute.

Juridical Courts

From what has been said thus far, one can clearly see that courts
occupy the upper end of the justicial scale in terms of both formality and
juridicality. The tribunal is today typified by the International Court of
Justice. Its function, under Article 38(1) of the Statute of the Court is to
decide disputes in accordance with international law.[31] In this sense, the court

is the living symbol of the rule of law as opposed to the rule of force. It adds stability to the international legal process by providing a forum of permanence by whose jurisprudence the law is allowed to mature. Against this background of expectations, one must view the reality that the tribunal of which we speak now is clearly the greatest threat to nations jealous of sovereignty, thus is least likely to be either the chosen or most effective forum for the resolution of disputes involving international relations which, while resolvable by principles of law, have strong political overtones. This is best typified by Litvinov's axiom "only an angel could be unbiased in Russian affairs."

The facts belie the expectations. The expectations for the International Court were as follows:

> ...the First Committee ventures to foresee a significant role for the new Court in the international relations of the future. The judicial process will have a central place in the plans of the United Nations for the settlement of international disputes by peaceful means.... In establishing the International Court of Justice, the United Nations hold before a war-stricken world the beacons of Justice and Law and offer the possibility of substituting orderly judicial processes for the vicissitudes of war and the reign of brutal force.[32]

Between 1945 and 1970, 175 treaties conferred jurisdiction on the International Court of Justice. During that same period, proceedings were instituted in fifty cases, of which fourteen were requests for advisory opinions and only fifteen proceeded to judgment on the merits. Furthermore, analyses of the cases reveal that all but nine were between nations having no major ideological differences.[33] Thus while the court has many advantages, the statistics indicate a need for further investigation and improvement. The record is not impressive.

A number of proposals have been made to change the court. These proposals cover a wide range of ideas relating to procedure, the membership and competence of the court, the basis of jurisdiction, and other such areas. More will be said about these problems subsequently, but suffice it to say at this point that while some changes could be salutory, the basic problem with the International Court remains one of widespread lack of confidence. In adjudications before the court one party wins and one loses. In the world of international politics, particularly in the sphere of negotiations, it is not necessary for there to be a complete "winner" and a total "loser." Political compromise is possible so that face-saving becomes an alternative. Added to this fact is the fear that members of the court are suspected not to fairly reflect the interests of the nations parties to the dispute, or that they might indeed apply provisions of law which one or both parties might view as out of date or reflecting the interests of a special power group. The attitude of Canada with respect to its 100-mile pollution zone legislation is typical of this feeling. The enormous increase in the number of independent states in the last two decades has added much fuel to this fire, particularly with respect to law of the sea matters. The political shift is from an essentially West European–American orientation to a greatly varied international community approach; thus nations would seem at this point in history to be uncertain as to the path of international jurisprudence, and, hence, the future of compulsory adjudication by a permanent court.

Another question often discussed relates to the respective role of the International Court of Justice and the possible roles of subsidiary international courts basing their jurisdiction upon geography or expertise. It has been variously suggested that regional courts be established, or that courts such as the proposed Law of the Sea Tribunal be created to deal with problems of special expertise. One must ask whether the creation of such tribunals would add to the cause of adjudication of international disputes or detract from it. On the one hand, creation of subsidiary instruments would detract from the prestige of the World Court. On the other, it may ultimately enhance its usefulness by creating an atmosphere where nations become more accustomed to the fact of international adjudication.

Clearly, with all of its faults, a permanent adjudicatory institution will continue to exist. Perhaps more than one will be required. It may be that adjudication by judges on questions of legal interpretation of treaties alone will not be sufficient, and that other types of machinery will be needed such as the World Equity Tribunal proposed by Clark and Sohn to deal with disputes not primarily of a legal nature.[34]

Relevant to the subject of specialized courts, the U.S. proposal for the settlement of disputes with relation to law of the sea would create a unique tribunal. Article 1 of that draft first provides for other means of settlement:

> In any dispute between the Contracting Parties relating to the interpretation or application of the present Convention, any party to the dispute may invite the other party or parties to the dispute to settle the dispute by direct negotiation, good offices, mediation, conciliation, arbitration, or through special procedures provided for by an international or regional organization.[35]

While such optional procedures are made available, any party to a dispute which is required by the terms of the Law of the Sea Convention to be submitted to compulsory settlement may at any time refer the dispute to the Law of the Sea Tribunal created by the articles. Thus the ultimate decision rests in the hands of the tribunal if other methods fail. While the tribunal would be established by the procedure provided for the International Court of Justice, it would also be assisted in the consideration of technical cases by four technical assessors sitting without right to vote. The specialized nature of the court is indicated by that provision, and by the provision in the draft which would provide the owner or operator of a vessel detained by any state the right to bring the question of detention before the tribunal to secure its prompt release, even though the dispute relating to the initial seizure of that vessel was being handled by some other procedure, such as arbitration.

The tribunal proposal is likely to fall upon dark days. Clearly, it suffers from the same suspicions attributed to compulsory tribunals generally. Yet to water the provisions down, such as the compromise of the Option Clause of Article 36 of the Statute of the International Court[36] would be to predispose the tribunal toward a record as undistinguished as the ICJ in terms of the numbers of significant disputes referred to it.

PROBLEMS RELATED TO DISPUTE SETTLEMENT

Several of the methods of international dispute settlement discussed in the preceding section have problems in common, and others have unique difficulties. In this section, there will be no attempt to exhaust these areas, but to highlight those which have a direct bearing on decisions with regard to fisheries disputes.

Structure—Number and Qualifications
Simply stated, a dispute may be referred for decision to a specially qualified individual (such as a particularly noted international lawyer); it could be referred to a commission consisting, as a minimum, of one national member from each party and an umpire agreed upon by both; or to a tribunal of three to five or more members with the issue decided by majority vote.

Certainly, single member settlements are most likely to be undertaken when devices such as good offices and mediation are utilized. In those cases, nothing is attempted except, in the one instance, a bringing together of the parties, and in the other, a bringing together accompanied by a suggested solution or solutions. A single mediator must be chosen on the basis of the authority he or it represents, or because of his or its reputation for justice and effectiveness. Thus, a single mediator could be a government or a private individual. The government has the advantage of added authority and the ability to provide a capable and numerically adequate staff. An individual, on the other hand, would normally have the advantage of flexibility, in that his representations would be personal ones, not subject as much to the need for internal discussion. Of course, both may occur when a government is selected for the reason that the parties know that the individual who will represent that government is one of high personal qualifications. The critical point is that the parties must agree and have confidence. Without this element, there will be no chance to bring together parties who have previously failed, themselves, to negotiate the issue. Because of the special nature of the selection of a mediator, it would be impossible to have a permanent mediator; but there could conceivably be

established an international panel of mediators who would be available to
parties to disputes. In general, however, it would be best to leave maximum
flexibility in this regard. The problem here, of course, is that agreement on a
mediator may take considerable time, delaying the discussion of the dispute
on the merits.

　　When disputes are considered to be more appropriate for a commission of inquiry, conciliation, arbitration, or judicial settlement, then
questions of numbers and selection become more pertinent. No single rule can
be applied that will satisfy the needs of each of these devices. In selecting members of a commission of inquiry, for example, it may be more important to
select members because of their special expertise in the area under dispute
than for their judicial qualifications. In the *Dogger Bank* case, previously
mentioned,[37] such was the situation. Because the problem was a naval one,
admirals were selected to constitute the commission. This is normally appropriate because most cases submitted to such commissions are normally not
complicated, there are no problems of law involved, and there are no evidentiary
decisions to be made regarding burden of proof. Such panels can normally be
selected and appointed with relative ease and speed, since no matters of high
national interest are involved.

　　When what is required is more than advice, more than fact-finding,
then the selection problem becomes of more importance. Thus in the areas of
arbitration and judicial settlement, where the tribunal is expected to rule on the
dispute, putting an end to it without appeal, the designation of the members of
the tribunal tends to be more discriminatory. Arbitration tribunals have been,
in the past, selected a number of ways. One common method is for each party
to the dispute to name one national, and the third, the chairman, is then selected
by agreement between the two. One of the problems here, of course, arises
when the parties fail to agree on the third member. One mechanism to deal with
this problem is that set forth in the Hague Convention of 1907:

> Each party appoints two Arbitrators, of whom one only can be its
> national or chosen from among the persons selected by it as mem-
> bers of the Permanent Court. These Arbitrators together choose an
> Umpire.
>
> If the votes are equally divided, the Umpire is entrusted to a third
> Power, selected by the parties by common accord.
>
> If an agreement is not arrived at on this subject, each party selects a
> different Power, and the choice of the Umpire is made in concert by
> the Powers thus selected.
>
> If, within two months' time, these two Powers cannot come to an
> agreement, each of them presents two candidates taken from the

list of members of the Permanent Court, exclusive of the members selected by the parties and not being nationals of either of them. Drawing lots determines which of the candidates thus presented shall be Umpire.[38]

Of course, this formula is only one of many procedures which have been built into substantive international agreements, but it is typical of the kind of mechanism that is normally used. Note also, that this particular tribunal is designed to have five, instead of three, members. The larger number is normally preferred because it allows for more views to be focused on the problem. However, where five are selected, it would normally be better form to permit each party to select one member, and then the other three be nonnationals selected by agreement. The advantage of this mechanism is that the three-man panels place the umpire, or chairman, in a difficult position. It is clear that the nationals will adopt their own state's view. Thus the full weight of deciding falls squarely upon the umpire. The addition of two other nonnationals to the procedure relieves him of a great deal of this weight, particularly if the tribunal is designed so that they will also not be from the same state as he. A further worry is how to bring the selection to an end within a reasonable period of time. If no safeguards are provided, a party to a dispute might accomplish his ends by simply refusing to cooperate; thus the clause in the Hague Convention limiting the first selection phase to two months' time. One of the most elaborate formulae devised for the selection of arbitrators is to be found in the Pact of Bogota, and it is reproduced in the footnotes for reference purposes.[39]

 Much has been said about whether there should be permanence in arbitration. A permanent council has the advantage of independence, impartiality, and prestige. Over the years it also tends to develop expertise in various areas. Of course, permanence and impartiality do not necessarily go hand in hand, and conflicts could arise in given cases where permanent members may have a direct or indirect interest. Expertise is another problem that has always haunted the processes. As previously mentioned, it is usually to be desired in fact-finding. How, on the other hand, does one assure the necessary degree of expertise in a permanent tribunal? One way is to permit the parties to replace their national representatives with someone having special expertise regarding a particular dispute. Another, as has been proposed in the United States' draft articles for a Law of the Sea Tribunal, is to provide the court with a panel of expert assessors who do not vote. Clearly some disputes would be better handled if the panel were not selected from permanent nominees, and where the *ad hoc* approach has been deemed appropriate, it has been utilized.

 Whether the arbitral tribunal is permanent or not, the important point is the desire of the parties with regard to picking the most objective and best qualified individuals. When this is emphasized, the chances of fair results are enhanced. For as Frankfurter once said:

> The fact is that on the whole judges do lay aside private views in
> discharging their judicial functions. This is achieved through train-
> ing, professional habits, self-discipline and that fortunate alchemy
> by which men are loyal to the obligations with which they are
> entrusted.[40]

The constitution of international courts, being of the highest order of dispute
settlement, tends toward the greatest degree of permanence. The selection of
the members of the ICJ, for example, is meticulously provided for in the Statute
of the International Court of Justice. The fifteen members are elected by the
General Assembly and the Security Council of the United Nations from a list of
persons nominated by the national groups in the Permanent Court of Arbitration. The term of office is nine years (with eligibility for re-election), and in
making a selection, not only the qualifications of the individual and his nationality must be considered, but representation must be assured among the major
legal systems of the world.[41] This system, of course, is a complex one, and
represents a certain amount of politicizing of the court. The political roles of
the court have frequently been debated, but, if Frankfurter is to be believed,
this problem may not in reality be so worrisome. At least, the problem is
different only in degree rather than in kind than faced by attorneys appearing
before domestic tribunals. As Jessup once pointed out:

> It is inevitable that attorneys for litigants should consider the
> composition of a court before which they may argue a case. In
> domestic practice there may be a choice of forum and in such
> instances the known point of view of a judge will be an influential
> factor in the choice. A skillful pleader may study all previous
> judicial pronouncements of a particular judge on a particular field
> or question of law to frame arguments designed to strike a respon-
> sive cord....[42]

Yet, to the sadness of some litigants, this process has never been considered a
foolproof approach to international jurisprudence. For it is amply clear that if
one attorney utilizes the device, the other shall also, and yet but one realizes
the victory.

To review this section, then, the problems which have been high-
lighted involve the number, the permanence, the expertise and qualifications of
those engaging in dispute settlement.

Identification of Substantive and
Procedural Matters

Here I shall deal with matters referred to earlier in the paper. It is
important that the dispute settlement mechanism clearly understand its correct
role. As was earlier noted, failure to derive this understanding could lead to

incorrect selection of the members. For example, it may be that admirals should not be selected for a commission of inquiry if that commission is also charged with a finding of fault.

If negotiation is the specific device under consideration, then matters dealing with rules to be applied have little relevance, for negotiators establish their own rules in the process. They can limit themselves to findings of fact, to applying law, or they can leave the judicial and enter the political realm by creating new law. Conciliation, on the other hand, is more restricted. If one views only the structure of a conciliation commission, then the process appears to be juridical. But if one focuses on the *objective* of conciliation, the production of a compromise, then the rule of law takes on less significance. In fact, conciliation is a mechanism for setting aside or modifying the rule of law in specific instances, although it is certain that the commission in doing so takes the rule into consideration. In that sense, in the Institute of International Law in 1961, some members propounded the theory that conciliation is a quasi-judicial procedure; i.e., the conciliation commission should set forth the facts and the applicable law before conciliating.

In the fields of binding settlement, an understanding of the procedural and substantive limitations on respective bodies becomes essential. A great deal of history supports the role of the binding arbiter. Clearly, if arbitration is to maintain a degree of viability among the many possible mechanisms, it must provide advantages and play roles different from them. While we take note of the flexibility of conciliation, we see that arbitration operates under more strictures.

What rules of law are to be applied by an arbitral tribunal? At an early stage Grotius distinguished between arbitration and judicial settlement; quoting Seneca:

> The condition of a good case seems to be better if it is referred to a judge rather than to an arbitrator; for the rules of law apply to the former and set certain limits, which he may not pass. In the case of the arbitrator, a religious scrupulousness, free and unchecked by restraints, can both take away and add to, and direct the decision not as the law or justice advises, but as humanity and pity move.[43]

The Hague Conventions for the Pacific Settlement of Disputes provided that international arbitration would settle differences "on the basis of respect for law," but added that arbitration is recognized as "the most equitable means of settling disputes which diplomacy has failed to settle." Read together, these provisions have been interpreted to mean that the arbitral body should decide in accordance with equity, that is, *ex aequo et bono*, in cases where it finds positive rules of international law to be lacking. The exact content of *ex aequo et bono* sometimes seems elusive. It is not the same as the term "equity" as that term is generally understood in Anglo-American jurisprudence. Yet it is not a complete invitation to arbitrariness. It is rather the body of general principles of justice as distinguished from any particular system of jurisprudence or the municipal law of any state. Thus if the principle is applied, the natural consequence is that the tribunal is deciding what is essentially a nonlegal dispute, for there is no specific rule of international law guiding the settlement. Yet it would be going too far to say that the dispute is purely political, since it is solvable by deriving from the general body of justice certain principles which in fairness ought to be applied. Others have viewed the application of the principle as an exercise of a legislative function, since it enables a tribunal to create nonexisting rules for the solution of a dispute. Viewed in that light, the power ought to be applied sparingly, and states should hesitate in entering into permanent arrangements giving arbitral tribunals such power, for to do so would be to create permanent legislative powers binding upon the parties. Perhaps such arrangements are best reserved to the settlement of specific *ad hoc* disputes wherein the parties agree, because of the nature of the dispute, motivated by the desire for compromise, to bestow the additional authority upon the panel.

At the top rung of the dispute settlement structure sits the International Court of Justice. Its function is to decide disputes in accordance with international law. This is clearly set forth in Article 38 of the Statute of the International Court of Justice:

1. The Court, whose function is to decide in accordance with international law such disputes as are submitted to it shall apply:

   a. international conventions, whether general or particular, establishing rules expressly recognized by the contesting states;

   b. international custom, as evidence of a general practice accepted as law;

   c. the general principles of law recognized by civilized nations;

   d. subject to the provisions of Article 59, judicial decisions and the teachings of the most highly qualified publicists of the various nations, as subsidiary means for the determination of rules of law.

2. This provision shall not prejudice the power of the Court to decide a case *ex aequo et bono*, if the parties agree thereto.

The United States' draft articles on compulsory dispute settlement, which propose a new Law of the Sea Tribunal, are silent on this issue. However, they do specify that the tribunal shall have jurisdiction over all disputes submitted to it as provided for in the agreement with reference to its interpretation or application and reference is made to a statute. Further, the method of election is the same as for the International Court of Justice, and it can be inferred that the tribunal shall have the same general role as the ICJ, except that it shall be limited to law of the sea matters. This is further reinforced by the provision that permits parties to utilize any of the other previously discussed mechanisms before resort to the tribunal. Hence, the tribunal is apparently viewed in historical perspective. This matter, as well as the question whether the tribunal may, if the parties agree, resort to *ex aequo et bono* is presumably left to the statute yet to be drafted. With regard to the latter, however, it is of interest that neither the Permanent Court of International Justice nor the ICJ have ever been called upon to render a decision *ex aequo et bono*.

Courts and tribunals of the nature of the ICJ, then, represent the culmination of international jurisprudence. Their presence signifies the existence and flexibility of rules of international law, and their permanence provides both a ready forum for dispute settlement and the cosmetic effect of advertising the desirability of the rule of law as opposed to the rule of force.

THE SETTLEMENT OF FISHERIES DISPUTES

In this section, the term "fisheries disputes" is intended to include the broad range of problems which may be faced by participants in international fisheries. Thus comment will be made concerning political or legislative mechanisms to deal with questions of change, fact-finding, and adjudication of individual disputes. Furthermore, as Christy has noted, disputes involving fisheries can be further subdivided into those over management decisions and those where the issue is the allocation or distribution of the benefits of fishing.

There are few mechanisms which have been formalized for fisheries dispute settlement between nations. The only notable multilateral agreement in force containing dispute settlement procedures is the Geneva Convention of Fishing and Conservation of the Living Resources of the High Seas of 1958.[44] The machinery which it establishes consists of a combination of direct negotiation and compulsory arbitration. Article 4 states:

> If the nationals of two or more States are engaged in fishing the same stock or stocks of fish or other living marine resources in any area or areas of the high seas, these States shall, at the request of any of them, enter into negotiations with a view to prescribing by agreement for their nationals the necessary measures for the conservation of the living resources affected.

Article 6(3) also provides for negotiations where a national of one or more states is fishing in an area adjacent to a coastal state where conservation is in issue. Article 6(4) calls for negotiations where there is disagreement concerning conservation measures in the high seas adjacent to a coastal state. Thus, in matters involving conservation, that is, matters of management, the device selected is direct negotiation. But if negotiation fails, then Article 9 is brought to bear:

> Any dispute which may arise between States under Articles 4, 5, 6, 7 and 8 shall, at the request of any of the parties, be submitted for settlement to a special Commission of five members, unless the parties agree to seek a solution by another method of peaceful settlement, as provided for in Article 33 of the Charter of the United Nations.

Article 11 makes decisions of the commission binding on the states concerned, except that such decisions can be revised if substantial factual changes, relating to the basis of the decision, take place. So far as can be ascertained, no arbitrations have taken place under the 1958 Convention, nor is it likely that there will be. One reason is probably the requirement of Article 10(1) (a) that in deciding disputes under Articles 4, 5, 6, and 8, the commission shall assure that measures agreed upon do not discriminate, in form or fact, against fishermen of other states. While the convention speaks in terms of conservation, it is clear that this is a veiled term of reference which is in fact dealing with the distribution of the resource. Thus a nondiscrimination clause would rest uneasily with states interested in increasing their own share of a limited resource at the expense of other states. In short, states are reluctant to submit economic interests to specialized commissions of biologists for binding arbitration. It is probably for the same reason that many major fishing nations, such as Japan, have never become signatories to the convention.

The European Fisheries Convention of 1964 is a like example of the utilization of binding arbitration.[45] That convention establishes the right of a participating coastal state to establish fisheries regulations to a distance of twelve miles.[46] It also protects special historical rights for the fishing vessels of contracting parties which habitually fished in regulated areas between January 1, 1953, and December 31, 1962. Article 13 provides:

> Unless the parties agree to seek a solution by another method of peaceful settlement, any dispute which may arise between the Contracting Parties concerning the interpretation or application of the present Convention shall at the request of any of the parties be submitted to arbitration in accordance with the provisions of Annex II to the present Convention.[47]

This convention, unlike the 1958 agreement, creates a list of permanent arbitrators, five from each signatory, to serve for terms of six years. A panel for a specific dispute consists of five members, each party naming one, and the remainder to be chosen by agreement between the parties from nationals of third states on the list. If these nominations are not made within one month from when the arbitration was first requested, the choices are made by the president of the ICJ from among nationals of states members of OECD, preferably, but not necessarily, from the list of arbitrators submitted by the contracting parties. There is no readily available information on whether this provision has fared any better than the mechanism established by the Geneva Convention, but an educated guess is that it has not. It suffers from the same deficiencies. However, it does provide for a degree of permanency lacking in the earlier document, making it possible for the parties to move more expeditiously toward a settlement since, presumably, the credentials of the list of arbitrators will be known, and thus a panel could be more quickly selected.

The general pattern of these multilateral conventions, which are given only as examples, is repeated in Article IX of the United States' revised draft articles on fisheries submitted to Subcommittee II of the Seabeds Committee on August 4, 1972. This article provides for a special "commission" of five members to be established at the request of any party to a dispute arising under the articles, unless the parties agree to seek a solution by another method of peaceful settlement provided for in Article 33 of the Charter of the United Nations. Members of the commission would be named by agreement among the parties within two months' time, but, failing agreement, they would be named by the secretary general of the United Nations "from amongst well-qualified persons being nationals of States not involved in the dispute and specializing in legal, administrative or scientific questions relating to fisheries, depending upon the nature of the dispute to be settled." The question of nationality is not well handled by this Article. In paragraph A, no mention is made of nationality in that the members should be "named by agreement," but as noted above, nationality is mentioned where the appointment is to be made by the secretary general. Paragraph B, however, states:

> Any State party to proceedings under these articles shall have the right to name one of the nationals to sit with the special commission, with the right to participate fully in the proceedings on the same footing as a member of the commission but without the right to vote or to take part in the writing of the commission's decision.

Thus one may conclude that the five-member commission must be appointed solely from third-party states, although this fact could have been more clearly elucidated.

The decisions of the commission, like the tribunals previously mentioned, are binding. Furthermore, the panel has the power to apply interim measures relating to conservation and "other measures" during the pendancy of the dispute, although there is no clarification as to what these other measures might be.

The draft language referring to persons specialized in legal, administrative, or scientific matters could indicate an attempt to avoid criticism that the commission was not competent to deal with matters affecting the economic interests of the disputing parties. It could also indicate an intent that the panel could exercise a wide range of functions including fact-finding and the interpretation and application of law. The indefinite nature of the language in this regard seems to suggest that no deliberate thought was given to the precise role of the commission, and the scope of its intended powers. Because there is no appeal from the commission, one can deduce that it is a tribunal with juridical powers. If so, it should have been more carefully structured.

A less effective device, but an attempt to deal with dispute settlement along different lines, is set up in the North Pacific Fur Seal Convention of 1957.[48] In that convention, it was specified that if any party considers that the obligations undertaken are not being carried out, the parties should consult on the need for and the nature of remedial measures, and failure to agree with regard to these measures could be cause for an aggrieved party to give notice of intention to terminate the convention. If this occurs, the convention terminates as to *all* parties nine months from the date of notice.

Some fisheries conventions do not deal with the question of dispute settlement *per se*. They do not establish arbitral or judicial mechanisms, but rather they are designed to shed light on problems in order that disputes can be minimized. The convention establishing the Inter-American Tropical Tuna Commission is typical of this type of treaty.[49] This convention establishes a commission to gather and interpret factual information to facilitate maintaining the maximum sustained catches of yellowfin and skipjack tuna, and other fish of concern, such as anchovetta. The major objective of the commission is to make studies and to:

> Recommend from time to time, on the basis of scientific investigations, proposals for joint action by the High Contracting Parties designed to keep the population of fishes covered by this Convention at those levels of abundance which will permit the maximum sustained catch.[50]

Thus the commission has duties and responsibilities which could be classified as quasi-legislative.

The International Court of Justice has played a role in fisheries disputes. Aside from the recent dispute between the United Kingdom and Iceland relating to Iceland's refusal to accept the jurisdiction of the court pursuant to their 1961 bilateral fisheries agreement, the most important single decision is the *Anglo–Norwegian Fisheries Case*.[51] This case established the general principle that it is not contrary to international law to draw straight baselines for use in measuring the breadth of the territorial sea where the coastline is deeply indented or fringed with islands. Keeping in mind the earlier discussion, one can easily see that this was a natural case for treatment by a judicial tribunal. It involved matters of high national importance, and it was decided by the application and interpretation of principles of international law. Conflicts of this order and structure are extremely appropriate for settlement by the court.

Despite this decision, however, the pattern that emerges from the treaties of record is one of binding arbitration. At least that seems to be the general preference of those who negotiated the treaties. However, the lack of evidence of sustained use of these mechanisms casts doubt on how much confidence either governments or individuals involved in fishing have in them. If this lack of confidence is real, what needs to be done for the future is to analyze the special requirements of fisheries disputes in terms of setting out alternatives by which these disputes might most expeditiously be resolved.

Among the more important problems that seem to confront the fisherman is the need for prompt relief. If the dispute involves questions regarding, for example, the setting of quotas, then the time factor may not be so critical. If, however, the dispute arises from the seizure of a vessel and cargo, the act of relief needed is one of urgency. While the ultimate propriety of the seizure may be left to future hearings on the merits, an immediate adjudication is necessary to obtain the release of the vessel under bond or an equivalent guarantee, sufficient to cover assessed penalties, should that be the final outcome.

The United States' draft articles for dispute settlement provide such a device. Article 8 states that:

1. The Tribunal shall expeditiously handle disputes which are of an urgent character and shall in appropriate cases issue binding interim orders for the purpose of minimizing damage to any party pending final adjudication. The Tribunal may also take such binding interim action in cases which have been submitted to arbitration....

2. The owner or operator of any vessel detained by any State shall have the right to bring the question of the detention of the vessel before the Tribunal in order to secure its prompt release in accordance with the applicable provisions of this Convention, without prejudice to the merits of any case against the vessel.

Not only would the Law of the Sea Tribunal, sitting as a judicial body, have the power to exercise the normal equitable remedy of granting injunctive relief, but a private party would have the right to raise the question.

This latter provision is rare in a system of law which has traditionally maintained that the state is the only correct legal personality. The compromise of permitting the individual to seek temporary relief and allocating the question of the merits to more formalized procedure seems a wise one.

In selecting a dispute mechanism for fisheries, one might classify the various categories of disputes that might arise. It is here that the author is particularly weak, and is admittedly speculating. One could perceive that the following kinds of disputes, among others, might occur:

1. Disputes concerning disagreement over facts. These might include, for example, disagreements whether particular stocks were being overfished, or whether a particular vessel was or was not within a certain defined limit or had fished certain stocks.

2. Disputes over the allocation of resources. Such disputes are political in nature, and involve both a degree of fact-finding and an understanding of national economic interests.

3. Disputes over the law. This category might include, for example, the validity of a claim to a specific fisheries limit under international law, the validity of a baseline, or whether a particular stock was intended to be included within a certain agreement.

Disputes over facts are the least contentious and the most susceptible of solution. In many cases, if the facts are clearly ascertained and publicly known, the parties can thereafter negotiate their differences, as in the *Dogger Bank* case,[52] either directly or through mediation or conciliation. This is particularly true when the facts show a clear violation by one of the parties.

Disputes over allocation of resources are essentially political, which is to say that they are solved by negotiating the economic interests of the parties. While each state will surely seek to maximize its share of the attainable resource at the expense of the others, an understanding of the full economic implications of one choice as opposed to another provides an essential starting point in the bargaining process. Because allocation questions are essentially economic, failure to achieve an agreement cannot readily be rectified by reference to a legal tribunal. Thus in the broad area of dispute settlement, we cannot look to juridical means to solve this basic negotiating problem, but must rely on broad multilateral understandings achieved through hard bargaining.

Disputes concerning the application or interpretation of international law remain within the province of arbitration or judicial tribunals. As previously noted, the latter normally is reserved for questions of high national interest involving interpretation of law by the rules of law. The former is more flexible, allowing other than the strict rule of law of any particular system of law to control. It has, in the eyes of lawyers, lent itself more readily to disputes of a commercial nature, both domestically and internationally. Where agreements have come upon a sticking point, it usually is the question of compulsion. And we return to where we began. Compulsory settlement is the *sine qua non* for reliability. Yet a widespread lack of confidence in international procedures has made the acceptance of such a procedure unlikely. How does one overcome this basic dilemma? If one is to assume that the states assembled at Caracas will not agree to surrender any portion of their sovereign rights to a superior tribunal, is all lost? It has been suggested that the question ought not to be nailed down with no avenue of future reconsideration. Perhaps it would be possible through the statute which would have to accompany any basic draft articles to specify certain areas in which nations could agree on compulsory settlement. Admittedly the negotiating process would eliminate most questions of any significance, but if nations would then make a conscientious effort to try to utilize the same mechanism for more and more kinds of disputes on a voluntary basis, then confidence would slowly grow as the competence of the court or tribunal became recognized through the body of law that would be generated. This might further be assisted if the court or tribunal would sit in panels dealing with specialized or regionalized problems, so that parties having a particular kind of problem, or working within a given region, might be further inspired to utilize the existing machinery. A long history of successful resolution of relatively minor disputes would increase the confidence of states to the point where compulsory settlement of all fisheries disputes might be a more realistic goal.

The question remains whether fisheries disputes would best be settled by arbitration or by court action. The underlying problems are, as previously mentioned, those of permanence, expertise, speed, and impartiality. The mechanism selected must be flexible enough to provide a maximum degree of all of these. A slight modification of the present U.S. draft may work. The expert assessors provided to the tribunal could double as a commission of inquiry if it appeared after initial investigation that the matter might be clarified by fact-finding. Furthermore, it would seem wise to explore other methods of selecting judges than that utilized by the ICJ. The primary need in the selection process is obtaining men of impeccable qualifications in the law of the sea, and it is not immediately clear that the General Assembly and the Security Council are the appropriate bodies to decide this issue. Finally, it would seem that, for the time being, states will have to be permitted to reserve to themselves matters of high national interest, at least until the competence of the tribunal is so overwhelmingly recognized that the question of compulsion becomes moot. Emergency procedures for vessel release and quick access to the tribunal are also essential features.

Ad hoc arbitration, of course, should always be reserved for those cases where the parties consider it appropriate. This is always available, and in no way detracts from the establishment of a permanent tribunal.

This paper has attempted to set down some of the considerations and issues relevant to discussion at this meeting, and is in no way considered exhaustive or authoritarian. If it stimulates discussion, then its purpose has been served.

Notes to Chapter 12

1. Lauterpacht, *The Function of Law in the International Community* (1933), p. B2.
2. Scott, *The Hague Peace Conferences of 1899 and 1907*, Vol. II (1909), p. 327.
3. This represents the original language of the articles, which were subsequently amended. For a discussion, see Lauterpacht, *supra* note 1, at 32.
4. For essays on the general operation of the Court, see Jessup, *The Price of International Justice* (1971); Allott, "The International Court of Justice," in *International Disputes: The Legal Aspects* (1972), p. 128 et seq.
5. Lauterpacht, *supra* note 1, at 17.
6. Cot, *International Conciliation* (1968), at 1.
7. *Systematic Survey of Treaties for the Pacific Settlement of Disputes 1928-1948* (1948), p. 557; 110 League Treaty Series 113.
8. Sohn, *The Function of International Arbitration Today* (1963), p. 15.
9. *Ibid.*, at 16.
10. *Ibid.*, at 17.
11. *Systematic Survey, supra* note 7, at 1108; 188 League Treaty Series 75.
12. *International Disputes, supra* note 4, at 85.
13. *Ibid.*, at 92.
14. For further consideration of methods of fact-finding within the United Nations, see *International Disputes, supra* note 4, at 172.
15. Scott, *supra* note 2, at 315.
16. Scott, *Hague Court Reports* (1916), p. 404.
17. Scott, *Hague Court Reports* (2d) (1932), p. 135. See also 16 *AJIL* 485 (1922).
18. Cot, *supra* note 6, at 8.
19. *Annuaire de l'Institut de Droit International*, 49 (1961) Vol. II, p. 195.
20. Oppenheim, *International Law*, 7th ed., Vol. II, pp. 12-20.
21. General Assembly, Official Records, 3d Session, Pt. I, *Resolutions*, pp. 21-25.
22. See Lauterpacht: *The Function of Law in the International Community*, pp. 260-69.
23. Cot, *International Conciliation*, at 181. In his footnote the author gives an example:

This is borne out by the statement made by Mr. Black in the affair of the City of Tokyo 5% Loan: The very fact that the parties once agreed on a compromise formula makes it inappropriate and inadvisable for me, as a Conciliator, either to attempt to choose between the view of the French Cour de Cassation and the Masse, on the one hand, and that of the Japanese Supreme Court and the Metropolis, on the other hand, or to undertake to interpret the Bonds myself... To follow any of these approaches would not be in the interest of the parties who have asked me to assist them in the settlement of their disputes."

24. Scott, *supra* note 2, at 89.
25. *Ibid.*, at 325, Article 37.
26. Article 33, previously referred to, states in full:

  1. The parties to any dispute, the continuance of which is likely to endanger the maintenance of international peace and security, shall, first of all, seek a solution by negotiation, enquiry, mediation, conciliation, arbitration, judicial settlement, resort to regional agencies or arrangements, or other peaceful means of their own choice.

27. Article 37:

  1. Should the parties to a dispute of the nature referred to in Article 33 fail to settle it by the means indicated in that Article, they should refer it to the Security Council.

28. Sohn, *supra* note 8, at 22.
29. It is interesting to note that the American delegate to the Hague Peace Conference of 1907 observed that the Permanent Court of Arbitration:

  ...is not permanent because it is not composed of permanent judges; it is not accessible because it has to be formed for each individual case; finally it is not a court, because it is not composed of judges. [Hague Court Reports, 1st Series, 1916, p. 1.]

30. *International Disputes, supra* note 4, at 115.
31. Article 38 defines the order in which authority is to be looked to in the consideration of disputes of international law referred to the Court. This provision is referred to later in the text.
32. UNCIO, Vol. 13, p. 393, Report of the Rapporteur of the First Committee.
33. *International Disputes, supra* note 4, at 134.

34.   Clark and Sohn, *World Peace Through Law: Two Alternative Plans* (1966), p. 338.

35.   Draft Articles tabled by the U.S. on August 22, 1973, before the Plenary Session of the Seabed Committee.   See, also, the introductory speech by Ambassador John R. Stevenson as reported in the State Department press release of that same date.

36.   The Optional Clause, contained in Article 36(3), permits states to make declarations with regard to the jurisdiction of the ICJ either unconditional, on the condition of reciprocity, or for a certain time.

37.   Text referred to in note 16.

38.   Article 45.   Scott, *supra* note 2, at 331.

39.   Article XL:

1.   Within a period of two months after notification of the decision of the Court in the case provided for in Article XXXV (i.e., when the Court declared itself to be without jurisdiction), each party shall name one arbiter of recognized competence in questions of international law and of the highest integrity, and shall transmit the designation to the Council of the Organization.   At the same time, each party shall present to the Council a list of ten jurists chosen from among those on the general panel of members of the Permanent Court of Arbitration of the Hague who do not belong to its national group and who are willing to be members of the Arbitral Tribunal.

2.   The Council of the Organization shall, within the month following the presentation of the lists, proceed to establish the Arbitral Tribunal in the following manner:

a.   If the lists presented by the parties contain three names in common, such persons, together with the two directly named by the parties, shall constitute the Arbitral Tribunal;

b.   In case these lists contain more than three names in common, the three arbiters needed to complete the Tribunal shall be selected by lot;

c.   In the circumstances envisaged in the two preceding clauses, the five arbiters designated shall choose one of their number as presiding officer;

d.   If the lists contain only two names in common, such candidates and the two arbiters directly selected by the parties shall by common agreement choose the fifth arbiter, who shall preside over the Tribunal.   The choice shall devolve upon a jurist on the aforesaid general panel of the Permanent Court of Arbitration of the Hague who has not been included in the lists drawn up by the parties;

e. If the lists contain only one name in common, that person shall be a member of the Tribunal, and another name shall be chosen by lot from among the eighteen jurists remaining on the above-mentioned lists. The presiding officer shall be elected in accordance with the procedure established in the preceding clause;

f. If the lists contain no names in common, one arbiter shall be chosen by lot from each of the lists; and the arbiter, who shall act as presiding officer, shall be chosen in the manner previously indicated;

g. If the four arbiters cannot agree upon a fifth arbiter within one month after the Council of the Organization has notified them of their appointment, each of them shall separately arrange the list of jurists in the order of their preference and, after comparison of the lists so formed, the person who first obtains a majority vote shall be declared elected.

Article XLV:

If one of the parties fails to designate its arbiter and present its list of candidates within the period provided for in Article XL, the other party shall have the right to request the Council of the Organization to establish the Arbitral Tribunal. The Council shall immediately urge the delinquent party to fulfill its obligations within an additional period of fifteen days, after which time the Council shall establish the Tribunal....

40. *Public Utilities Commission v. Pollack*, 343 U.S. 451, 466 (1952).
41. See Chapter I, *Organization of the Court*, Articles 2 through 33.
42. Jessup, *supra* note 4, at 57.
43. *De Beneficiis*, II, vii, as quoted in Sohn, *supra* note 8, at 41.
44. Adopted by the United Nations Conference on the Law of the Sea, April 28, 1958 (U.N. Doc. A/CONF.13/L.54).
45. *International Legal Materials*, Vol. 3 (1964), p. 476.
46. *Ibid.*, Article 3.
47. *Ibid.*
48. TIAS 3948.
49. TIAS 2044.
50. *Ibid.*, Article II (5).
51. *United Kingdom v. Norway*, (1951) ICJ 116.
52. *Supra* note 16.

List of Participants

Paul L. C. Adam
Head of Fisheries Division, OECD
Paris, France

Lewis M. Alexander
Professor of Geography
University of Rhode Island
Kingston, Rhode Island

Richard B. Allen
Professor of Fisheries and
Marine Technology
University of Rhode Island
Kingston, Rhode Island

Raoul Andersen
Professor of Anthropology
Memorial University
St. John's, Newfoundland, Canada

Lee G. Anderson
Professor of Economics and
Marine Science
University of Miami
Coral Gables, Florida

J. C. Arnell
Consultant, Ministry of Finance
Hamilton, Bermuda

Bessie R. Barton
I.S.M., Principal Customs Officer
Hamilton, Bermuda

James S. Beckett
International Fisheries and
Marine Directorate
Ottawa, Ontario, Canada

Richard B. Bilder
Professor of Law
University of Wisconsin Law School
Madison, Wisconsin

William T. Burke
Professor of Law
University of Washington School of Law
Seattle, Washington

James Burnett-Herkes
Curator, Bermuda Government
Aquarium and Museum
Flatts, Bermuda

Marlene Christopher
I.S.M., Assistant Registrar General
Hamilton, Bermuda

Francis T. Christy, Jr.
Resources for the Future, Inc.
Washington, D.C.

Thomas A. Clingan, Jr.
Professor of Law
University of Miami
Coral Gables, Florida

Jacob J. Dykstra
President, Point Judith Fishermen's
Cooperative Association, Inc.
Narragansett, Rhode Island

Roger A. Farge
Master Mariner, Operations Manager
Gypsum Transportation, Ltd.
Hamilton, Bermuda

Dennis Farias
Chairman, Bermuda Commercial
Fishermen's Association
Hamilton, Bermuda

John King Gamble, Jr.
Executive Director, Law of
the Sea Institute
University of Rhode Island
Kingston, Rhode Island

Gordon R. Groves
Director, Department of
Agriculture and Fisheries
Paget, Bermuda

William C. Herrington
Staffordville, Connecticut

Douglas M. Johnston
Professor of Law
Dalhousie University
Halifax, Nova Scotia, Canada

Albert W. Koers
Professor of International Law
Institute of Public
International Law
University of Utrecht
Utrecht, The Netherlands

Austen Laing
Director-General, The British
Trawlers' Federation Ltd.
Yorkshore, England

J. L. McHugh
Professor of Marine Resources
Marine Sciences Research Center
State University of New York
Stony Brook, New York

Gloria McPhee
Minister of Education
Paget, Bermuda

Edward Miles
Professor of Marine Studies
University of Washington
Seattle, Washington

Byron F. Morris
Oceanographer, Bermuda
Biological Station
St. George, Bermuda

Alfred W. H. Needler
Executive Director
Huntsman Marine Laboratory
St. Andrews, New Brunswick, Canada

John J. Owens
Lawyer
Oxnard, California

Choon-ho Park
East Asian Legal Studies
Harvard Law School
Cambridge, Massachusetts

James A. Pearman
Barrister, Conyers Dill & Pearman
Hamilton, Bermuda

Giulio Pontecorvo
Professor of Economics
Columbia University Graduate
School of Business
New York, New York

Rudolph F. Richardson
Charter Fisherman
Harrington, Smiths', Bermuda

Kenneth E. Robinson
Chief Education Officer
Department of Education
Bermuda

Earle E. Seaton
Puisne Judge, Supreme Court
Hamilton, Bermuda

Gunnar G. Schram
Deputy Permanent Representative
of Iceland to the United Nations
New York, New York

Anita B. R. Smith, M. P.
Hamilton, Bermuda

John Stubbs
Surgeon
Pembroke, Bermuda

Frank Watlington
Oceanographer
Sofar, St. Davids, Bermuda

Paul A. Welling
Department of State
Washington, D.C.

Elizabeth Young
Writer
London, England

Richard Young
Attorney at Law
Van Homesville, New York

About the Editor

Giulio Pontecorvo received his A.B. and M.A. degrees from Dartmouth College and his Ph.D. (economics) from the University of California at Berkeley. He has taught at the University of California, the University of Colorado, the University of Washington and Bowdoin College. Since 1963 he has been at Columbia University where he is Professor of Economics in the Graduate School of Business. Currently he serves on the Ocean Policy Committee of the Ocean Affairs Board, the National Academy of Science, and on the Executive Board of the Law of the Sea Institute.